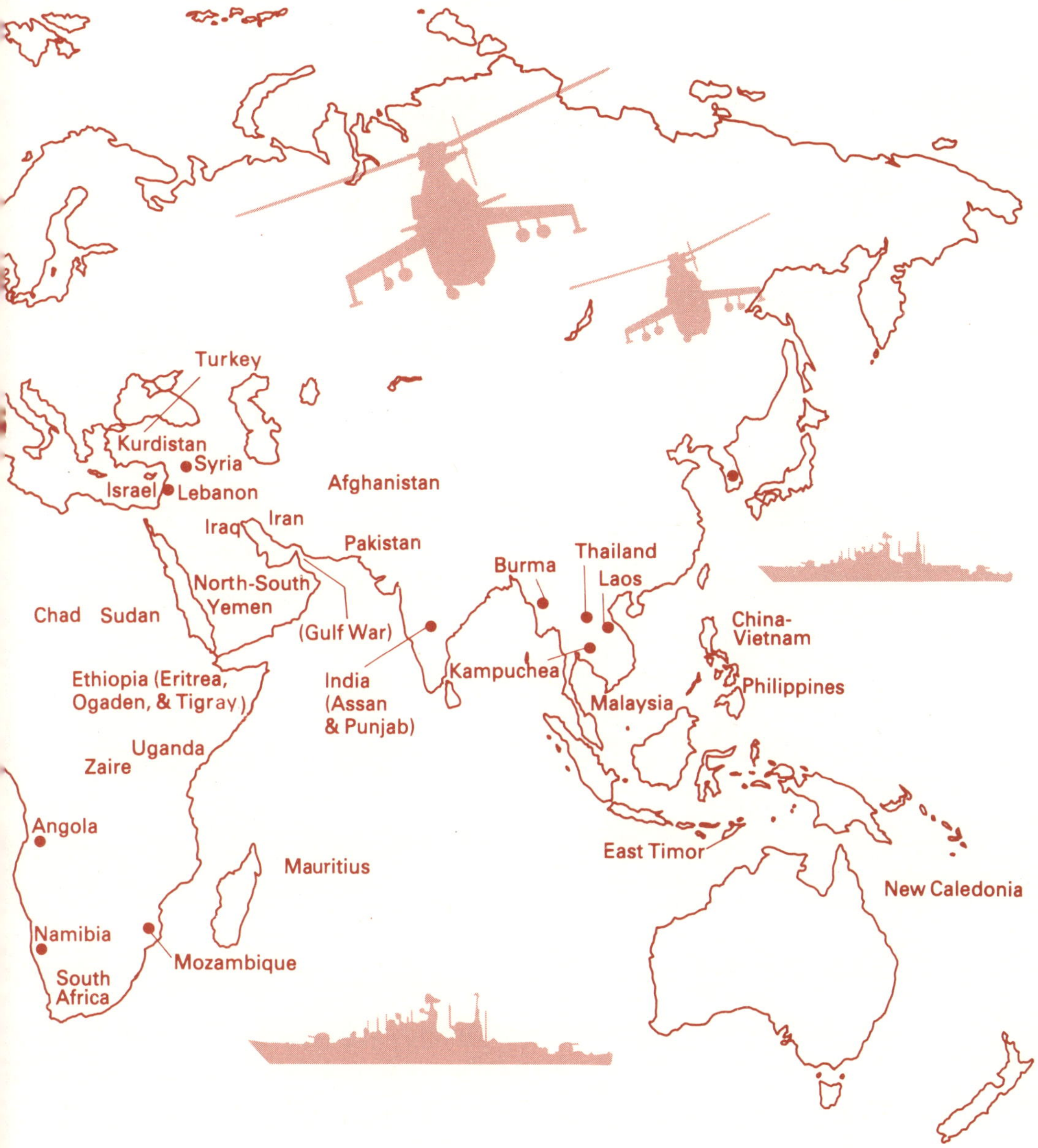
Turkey
Kurdistan
Syria
Israel
Lebanon
Afghanistan
Iraq
Iran
Pakistan
Thailand
Burma
Laos
North-South
Yemen
(Gulf War)
Chad
Sudan
China-
Vietnam
Ethiopia (Eritrea,
Ogaden, & Tigray)
India
(Assan
& Punjab)
Kampuchea
Philippines
Malaysia
Uganda
Zaire
Angola
Mauritius
East Timor
New Caledonia
Namibia
Mozambique
South
Africa

WAR ANNUAL 1

Titles of related interest from Brassey's

BELLAMY, C.
Red God of War: Soviet Artillery and Rocket Forces

COKER, C.
A Nation in Retreat?: Britain's Defence Commitment

LAFFIN, J.
Brassey's Battles: 3,500 Years of Conflict, Campaigns and Wars from A – Z.

MASON, R. A.
War in the Third Dimension

SIMPKIN, R.
Race to the Swift: Thoughts on Twenty-First Century Warfare

SLOAN, C.
Mine Warfare on Land

SWEETMAN, J.
Sword and Mace: Twentieth-Century Civil Military Relations in Britain

WAR ANNUAL 1

by

JOHN LAFFIN

BRASSEY'S DEFENCE PUBLISHERS
(a member of the Pergamon Group)

LONDON · OXFORD · WASHINGTON · NEW YORK · BEIJING
FRANKFURT · SÃO PAULO · SYDNEY · TOKYO · TORONTO

U.K. (Editorial)	Brassey's Defence Publishers, 24 Gray's Inn Road, London WC1X 8HR
(Orders)	Brassey's Defence Publishers, Headington Hill Hall, Oxford OX3 0BW, England
U.S.A. (Editorial)	Pergamon-Brassey's International Defense Publishers, 1340 Old Chain Bridge Road, McLean, Virginia 22101, U.S.A.
(Orders)	Pergamon Press, Maxwell House, Fairview Park, Elmsford, New York 10523, U.S.A.
PEOPLE'S REPUBLIC OF CHINA	Pergamon Press, Qianmen Hotel, Beijing, People's Republic of China
FEDERAL REPUBLIC OF GERMANY	Pergamon Press, Hammerweg 6, D-6242 Kronberg, Federal Republic of Germany
BRAZIL	Pergamon Editora, Rua Eça de Queiros, 346, CEP 04011, São Paulo, Brazil
AUSTRALIA	Pergamon-Brassey's Defence Publishers, P.O. Box 544, Potts Point, N.S.W. 2011, Australia
JAPAN	Pergamon Press, 8th Floor, Matsuoka Central Building, 1–7–1 Nishishinjuku, Shinjuku-ku, Tokyo 160, Japan
CANADA	Pergamon Press Canada, Suite 104, 150 Consumers Road, Willowdale, Ontario M2J 1P9, Canada

First edition 1986

Library of Congress Cataloging in Publication Data

Laffin, John.
War annual I.
1. Military history, Modern—20th century.
2. Military art and science—History—20th century.
I. Title. II. Title: War annual 1. III. Title: War annual one.
U42.L34 1986 355'.009'04 86–847

British Library Cataloguing in Publication Data

Laffin, John
War annual I.
1. War
I. Title
909.82 U21.2

ISBN 0–08–031211–X (Hardcover)
ISBN 0–08–031212–8 (Flexicover)

Printed in Great Britain by A. Wheaton & Co. Ltd. Exeter

Contents

Introduction

War is mankind's great preoccupation, and since 1939 there has not been a single day when world-wide peace prevailed. Quite possibly, fighting has raged somewhere in the world every day for centuries. Most people think about armed conflict in terms of the "great wars"—in modern times World Wars I and II. Other wars intrude from time to time because they attract world-wide attention, such as the war in Vietnam 1959-70.

Wars are memorable for a particular people at a particular time. The Falklands War of 1982 is likely to remain in the British consciousness for a generation—and as the Malvinas War it will stay in the Argentinians' mind. For the French the Falklands War was a minor conflict; for them the wars in Indo-China during the early 1950s and the Algerian War of the 1960s were the most significant in 40 years.

It is a simple if sad fact that judged by its frequency war is normal. For some peoples it is a noble activity, for others it remains the traditional and natural way of achieving some deeply desired ambition or aspiration—such as the acquisition of more territory or the imposition of an ideology or religion on other people. In our time independence has inspired as many wars as any other factor.

All the belligerent parties to a war consider their cause to be righteous. They are *for* or *against* something, such as a religion or political system. They fight for democracy or communism or nationalism—though the definitions of what these terms mean can differ to a bizarre degree.

This book is, I believe, the first attempt to describe all the wars in progress during an entire year. Without this knowledge it is hardly possible to assess and quantify the tensions and strains, grievances and hatreds that afflict such vast numbers of people.

I make no attempt to take sides in this book, and any comments I might make are those of the professional observer; when one side or another has been guilty of barbaric treatment of its enemy, then this must be said. It can be said here and now that very often *both* sides are guilty of brutal behaviour.

My purpose is to describe the background to each war, its course, conduct and conclusion—where a conclusion has been reached. Some wars since 1945 have been in progress for 20 years.

My sources are many and varied. During a lifetime as a writer on war I have built up hundreds of contacts in many countries; a good number of these contacts are in privileged or "sensitive" positions and must therefore remain confidential. I must say that I never accept an official statement about the course of a war without independent verification. Official communiqués almost invariably exaggerate enemy casualties and minimise their own, just as they justify the actions of their own leaders while deploring those of the enemy. As has been so cynically but truly said, "The first casualty in war is truth".

This annual is an attempt to report the truth.

The Wars — 1985 - 86

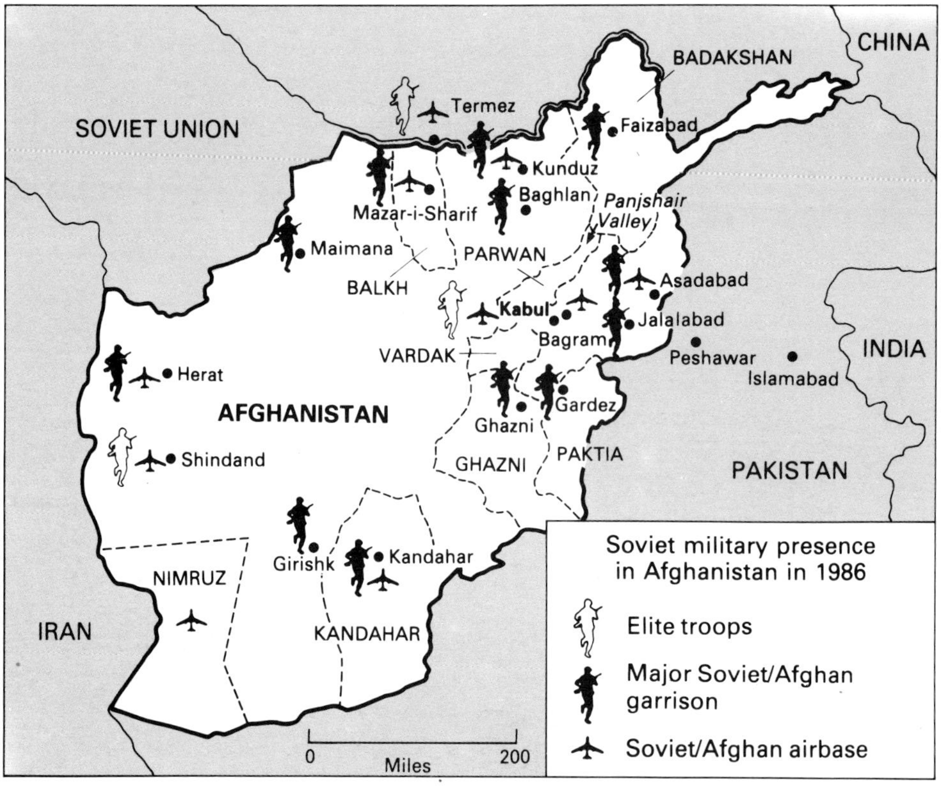

Soviet Invasion and Occupation of Afghanistan

Soviet Invasion and Occupation of Afghanistan

RUSSIAN TESTING GROUND

Background

The Soviet invasion of Afghanistan in December 1979 was carefully planned over the previous 2 years with Russian *agents provocateur* and pro-Soviet Afghanis preparing the ground by creating "incidents". Their purpose was to give President Hafizullah Amin an excuse to ask the Soviets, under a treaty of friendship, to send troops to assist the Afghan Army to maintain control.

During November 1979 the Russians sent small numbers of troops. After 3 weeks Soviet military strength had reached the level of about a brigade at each of three strategically important places—the air base at Bagram, 50 miles north of Kabul, the air base at Shendan in the west and the capital itself.

The invasion itself took place between 24 and 26 December when Soviet Army transports flew into Kabul all the elements of an airborne division of 6,000 troops. On the night of 27 December these troops attacked the radio and television station next to the US Embassy, the presidential palace and Duralaman palace where President Amin had just set up residence. Amin, who had himself deposed the previous president, was killed during the fighting. There is little doubt that he was murdered by the Russians who had decided that he was too indecisive and not a strong enough Party man for their purposes.

The Soviet high command had massed 50,000 troops, supported by tanks and helicopter gunships, on their side of the border. Simultaneously with the attacks in Kabul, the ground troops swept across the border in a three-pronged blitzkrieg-like invasion. A column crossed at Torghundi and took the provincial city of Herat, a second took Mazar-i-Sharif and the third drove rapidly down the Kunduz Valley, north of Kabul.

Babrak Kamal declared himself president and, as a Soviet puppet, ordered the Afghanistan Army to co-operate with the Soviet Army. This army of occupation was made up of 40% Tadzhik, 25% Uzbek and 25% Turkoman soldiers, all from Soviet Central Asia, the remainder being Russians. Since they could well become demoralised and even infected by the spirit of rebellion by having to fight people with whom they have racial links, it seemed that the Soviet command expected the take-over to be swift and decisive.

Indeed, it seemed that the country had become a Soviet satellite and that fighting had been minimal and was now over. The Russians claimed that they had saved Afghanistan from being subverted by the CIA and turned into an American base. Another Soviet propaganda line was that the US had collaborated with Pakistan, China and Egypt to carry out "anti-Soviet" propaganda in Afghanistan and had weakened the Afghan Army before a planned take-over of the country.

It was certainly true that large parts of the army had rebelled; desertions had cut its strength from 150,000 to 80,000. General Ivan Pavlovsky, chief of Soviet ground forces, had led a high-level delegation to investigate the unrest in Afghanistan.

The Soviet timing was good. The US administration of President Carter was distracted by the revolution and anti-American activity in Iran and was unlikely to intervene in Afghanistan. The discussion about the Salt II Treaty then in progress in the US Senate was also no obstacle to invasion: the Kremlin had already concluded that the arms treaty had no chance of winning Senate approval. Salt II no longer seemed important to the Russians because NATO had decided, in December 1979, to deploy cruise and Pershing II missiles in western Europe by the mid-1980s.

The Russian leaders had three strategic objectives.

1. To tighten their hold on the rebellious country. The tide of Islamic fervour, which had already shaken Iran, was now threatening Afghanistan. Unless checked, it might spread across the border into the Soviet Central Asian Republics and stir unrest among their Islamic populations of more than 50 million.
2. To gain influence in the long term over Pakistan, Iran and other Persian Gulf nations. The Soviet has a vested interest in securing influence in Iran. The prize in political, economic and military terms would be enormous. The Russians could turn off the oil tap for Western consumers almost at will when the oil shortage starts to bite, as expected, in the late 1980s.
3. By controlling Afghanistan, the Russians would be in a position to break through either Iran or Pakistan's Baluchistan province, reach the Indian Ocean and achieve the ancient Russian dream of a southern port.

Beyond any specific and immediate goals, the Soviet also intended their invasion of Afghanistan to demonstrate to Iran, Pakistan and China that "unruly" neighbours must expect punishment.

It is possible that the Russians were tempted to become increasingly adventurous at this time because the Pentagon had lost its clear-cut global strategic superiority.

Beyond all this, the Soviet message was addressed to the world, for all international relations depend on various nations' judgements about other nations' ability and willingness to use force. Regardless of the issues or moralities involved, a nation that occasionally asserts itself is often subsequently treated with greater care. The Soviet leaders were stating that they were prepared to use force.

The Soviet high command had calculated that by neutralising the bulk of the Afghan armed forces and by cowing any parts of it that might want to fight they could conquer Afghanistan. The planners realised that there would be some resistance from tribesmen in the hills, but they did not take it seriously because an Intelligence report stated that there was no unity among the tribes and that, in any case, they were armed with nothing more modern than 1945-type British Lee-Enfields. In under-estimating the Resistance movement the Russians made a serious miscalculation. By 1986 the Russians and their pro-Soviet Afghan allies had killed 300,000 Afghanis and driven 3.5 million of them out of their own country as refugees, but they had still not conquered Afghanistan and the conflict had become one of the most brutal and savage of the twentieth century.

The Afghan freedom fighters, the Mujahideen, are Muslims. This patrol was photographed at Friday prayers before setting off to attack a Soviet target. (Courtesy *Daily Express*.)

Fighters of the Mujahideen resistance organisation look down on the wrecked and deserted village of Domgon after a Soviet attack.

The Russians' failure to subdue the Afghani populace did not prevent them from building up a vast military infrastructure for the conduct and support of large-scale air operations, including major bases with long paved runways at Bagram, Kabul, Mazar-i-Sharif, Jalalabad, Herat, Shindand, Farah, Lashkar, Gah, Askargh and Kandahar.

During the period 1980-4 the Soviet armed forces and their Afghan regular army allies used the tactics of conventional warfare against the guerrillas—generally now known as the Mujahideen—who used their age-old hit-and-run methods of attack, with hill ambush as the main method of defence. In the large areas of wild mountains and hidden valleys such a conflict was bound to be protracted and inconclusive. The Soviet command believed during 1981 and 1982 that it could wear down the Resistance; instead, guerrilla groups proliferated until at least ninety were operating.

The War in 1985

During 1984 and increasingly in 1985 two factors made the war more bitter and intense.

(i) The Mujahideen received better and more continuous supplies of weapons mainly from the US, but also from Egypt, Saudi Arabia, Pakistan, Britain and China.

(ii) The Russians changed their strategy: Instead of merely trying to defeat the guerrillas, they aimed at driving them and their families out of Afghanistan. Also, they at last managed to give their troops a "cause". From the beginning of the war the Russian soldiers could see no reason to fight. They did not hate the Afghans and the Resistance reported that sometimes the soldiers chose to fire in the air rather than obey orders to shoot villagers. But the Mujahideen, increasingly under the influence of extremist mullahs, tortured captured Russians before killing them. Mutilated bodies were left on roads and paths to be found by horrified comrades. It became not uncommon for an officer to shoot himself when in imminent danger of being captured by the Mujahideen.

Exploiting this, the Soviet command introduced a programme to increase fear and hate and thus make more effective soldiers of the Russian troops.

The Helsinki Watch Committee, an independent human rights group, produced a report entitled *Tears, Blood and Cries, Human Rights in Afghanistan 1979-1985*. It describes terror tactics including torture and assassination used by both sides. The 212-page report devotes 172 pages to the Soviet and 16 pages to the Resistance, a disproportionately low figure.

The immediate American response in 1979 was to reinforce bases in the Indian Ocean and to designate the Gulf, Oman and Pakistan as possible areas of intervention, and to bring in a new partner, China. This stance was maintained into 1985-6.

While American military aid to anti-Communist forces in Nicaragua and Salvador received enormous publicity during 1985, the fact is that the CIA's aid to the Afghan rebels is the largest US covert operation since the Vietnam war era. Congress tripled the Reagan administration's initial request for the Afghan programme to about $250 million for the 1985-6 fiscal year. This is more than 80% of the CIA's annual expenditure for covert operations.

By the end of 1985 the American programme, which supplies weapons, ammunition, clothing and medical supplies and money for food, supported an estimated 200,000 full-time or part-time rebels. At times the flow has been so great that the supply line, which runs through Pakistan, could not readily absorb it. Despite the scale of military aid, US Intelligence officers cannot and do not control the operations of Resistance fighters.

During 1985 the Soviet President, Mikhail Gorbachev, three times warned President Zia of Pakistan to stop all aid from reaching the guerrillas or "face economic and military consequences". The Russians frequently threaten to use the "right of hot pursuit" against Islamic insurgents operating from Pakistan into Afghan territory.

During 1984 and 1985 the Russians carried out scores of ground and air space violations in Pakistani territory, killing more than 600 people and causing extensive damage. Pakistan received a total of $750 million in Soviet loans between 1981 and 1985, but aid has now been drastically cut.

The weapons and ammunition enable the Resistance to make more sustained and ambitious attacks than was possible with out-of-date rifles. During 1985 fighting was continual, with the Resistance fighters almost invariably taking the initiative. For instance, on 24 April guerrillas fired twenty rockets at Bagram air base and at about the same time attacked a Soviet post at Barandai Ianpanushir, killing thirty Russians. In another incident the Resistance fighters destroyed a pontoon bridge in Panjshair and attacked a Soviet convoy. They claimed to have destroyed several tanks and caused "very high" casualties.

Incidents in and around Kabul included a bomb explosion at the Ministry of Agriculture, rocket and bomb attacks on areas used mainly by Soviet and Afghan regime forces and officials, and attacks on the regime's army headquarters at Pol-i-charkhi to the east of the city and military positions. A number of Soviet and Afghan soldiers were killed or injured, including a Soviet general killed in the Microrayon district. The security perimeter extends for about 10 km from the city centre, with mainly Soviet troops stationed on high ground in a vain attempt to prevent rocket attacks.

The Panjshair Valley north of Kabul is the scene of revived activity. Helicopter-borne Soviet commandos tried to clear some of the side valleys, leading to 3 days of heavy fighting with Mujahideen forces occupying the valleys; 17 Soviet and 30 Afghan regime soldiers were killed. The Soviet aim was to seal off the southern approaches to the valley pending a large-scale offensive. At Kunduz, nearer to Afghanistan's northern border, 900 civilians were massacred by the Soviet Army in the second half of February 1985.

The highway running north to the Soviet frontier, and particularly the Salang tunnel, 125 km north of Kabul, remains a major Resistance target. Convoys are often attacked near the tunnel. During one week 300 vehicles, many carrying bombs, were destroyed and 400 Soviet/Afghan regime soldiers were killed.

West of Kabul, Soviet forces suffered another setback when they entered the Mayden Valley, only to be ambushed by Mujahideen from the Hazarajat. A large number of Soviet wounded were treated at a Kabul hospital as a result of the ensuing fire-fight.

In the south and east, 300 Afghan regime and Soviet soldiers were killed in attempts to relieve the garrison at Barikot, close to the border with Pakistan, which

The Soviet newsagency TASS is a major purveyor of propaganda about the "friendship" between Soviet soldiers and the Afghan people. This photograph showing Russian soldiers and co-operative Afghans was carefully contrived.

Afghan rebels in the preliminary stages of an attack on a Soviet post. The Russians have found no effective answer to the guerrillas' hit-and-run tactics. (Courtesy *Daily Express*.)

has been under heavy siege by the Mujahideen since mid-1984. Many of their tanks and vehicles were also destroyed. In later efforts to relieve Barikot, two Soviet companies were moved from Jalalabad to Hashkut in Kunar province, Soviet helicopters having earlier killed thirty civilians in raids nearby.

Revised and co-ordinated Soviet tactics against the Resistance during 1985-6 reflect the desperation of the Russians and their recognition that the Resistance is formidable. These tactics are based on constant use of airpower, airborne response, terrorism against villagers, on propaganda and "re-education" and the employment of élite forces, and attempts to cut the supply line.

Airpower

The pattern of air attacks became standard in 1985. First jets, operating in pairs, make sweeping circles along the mountain ridges, partly to intimidate but mainly for high-level observation to find guerrilla posts or patrols. Then come the helicopter gunships, two Mi-8 Hips leading, with Mi-24 Hinds following. Firing their multiple cannons and rockets, they try to smother guerrilla machine-gun fire.

Airborne Response

Because the Mujahideen sometimes capture and hold Soviet or Afghan regime posts, the Soviet Army has countered with rapid airborne response. For instance, guerrillas took three border posts on the Khyber pass. The Russians quickly sent in a full airborne brigade to retake these outposts.

Terrorism

Villages suspected of having harboured guerrillas are raided by Soviet troops, who slaughter livestock, set fire to granaries and level laboriously constructed soil terraces. Villages and crops near roads are methodically bombed. Most serious is the destruction of the irrigation system on which Afghanistan's agriculture is founded. Massacre is commonplace. Soviet commandos killed probably 1,000 people in two looting and burning raids in Qarghai district of eastern Afghanistan during April 1985. After the Resistance attack on Bagram on 24 April 1985, Soviet helicopters took off to pursue the guerrillas, could not find them and unloaded their bombs on villages in the area. During May, in Maidan Shehr, Wardak province, Russian troops seeking guerrillas killed 300 civilians in reprisals, which are commonplace.

Propaganda and "Re-education"

Late in 1985 the puppet government held a *loya jirga* (or grand assembly) for the first time since the invasion. A *loya jirga* is the traditional way for Afghan rulers to resolve complex political problems. Having managed to bribe some tribal leaders, the government hoped to persuade Afghans that the regime is not Russian and alien but has evolved out of Afghan society. The attempt failed. In the north the tribes that span the Afghan–Soviet border—the Uzbeks, Turkomans and Tajik—were fed propaganda that puts emphasis on their own culture. The main thrust of the campaign to pacify the country, however, is directed against the young throughout Afghanistan. The government publicises the work of children who have re-educated their "backward" parents. Higher education has been extended, and about 10,000 students a year are sent to the Soviet Union for training. From November 1984 large batches of children between 7 and 9 years of age have been sent to Russia for 10 years' schooling.

Élite Units in Action

When the Soviet command posted Spetsnaz units to Afghanistan it was more obvious than ever that the Resistance was being effective. The Spetsnaz is the Russian equivalent of the SAS and is the cream of the Soviet military élite. Using surprise tactics, small but highly trained Spetsnaz units have ambushed guerrillas in their own hills, a feat considered impossible by centuries of enemies. Even so, Dr. Najibullah, head of the Khad, the regime's secret police, admitted to an Indian journalist late in 1985 that only 35% of Afghan territory was under the regime's control.

Supply Line Disruption

The aim of many Soviet offensives near the Pakistani border is to establish forward positions and disrupt the flow of supplies. Soviet aircraft frequently cross the border, bombing in and around the Pakistani town of Arandu.

There is no foreseeable end to the war, and it might well become more intense. The guerrillas already have SA-7 ground-to-air missiles from Egypt, and the US has made the decision to give the guerrillas supplies of a new rapid-fire anti-aircraft cannon firing armour-piercing explosive shells. The decision is controversial. A US Intelligence officer said: "If helicopters start getting shot out of the sky with frequency we've got a problem. . . . A weapon like this could force the Soviets to become more indiscriminate in their use of force. They could do much more bombing".

The cost of each new cannon, plus transportation and initial supplies of ammunition, is about $1 million. Some estimates suggest that millions of dollars will have to be spent on ammunition each year.

The Soviet Union will protect its substantial investment. It is building up a vast military infrastructure in Afghanistan while preparing it as a potential strategic base for power projection throughout South-East Asia and the Middle East. An authoritative article in the Communist Party's own official daily *Pravda* in March 1985 gave a hint of the importance which Soviet leaders give to Afghanistan. *Pravda* said that Soviet troops stationed there are "defending Soviet borders from the approach of hostile imperialist and pro-imperialist forces".

This language indicates plans for expansion.

The Opposing Forces in 1985

Afghan

Army: 30,000. Nominally the army has 3 corps HQ, 11 infantry divisions, 3 armoured divisions, 1 mechanised infantry brigade, 2 mountain regiments, 1 artillery brigade and 3 commando regiments. All these units are under strength.

The army is poorly trained and ill equipped; its officer corps is split by factional political strife and is distrusted by the Soviet occupation forces. The tank corps, once the army's élite force, is virtually inoperative.

The Karmal regime uses every device to maintain the army's strength. These range from grants of money and land to soldiers willing to extend their period of service, to tough penalties for draft-evaders and use of boys aged 15–18. Recruitment is failing to meet the heavy losses caused by the fighting, by desertion and defection to the Mujahideen.

Air Force: 5,000, with 120 combat aircraft and 30 armed helicopters, mostly Mi-24s. There are 4 transport squadrons and 4 helicopter transport squadrons.

Paramilitary forces: About 30,000, comprising gendarmerie, border police, secret police, regional militias, Afghan Communist Party guards and "youth troops".

Soviet Forces in Afghanistan

Army: 115,000 at the end of 1985, but occasionally the number is boosted considerably and later reduced. About 12,000 of those men are members of the KGB and MVD. Significantly, most of the troops from Central Asia and all Muslim troops were gradually replaced by units from the western Soviet states, including some from Poland. The force comprises 1 HQ, 1 airborne division, 1 air assault brigade and 3 motor rifle divisions.

Air Force: About 300 jet fighters and 300 helicopters were stationed in Afghanistan late in 1985.

The senior commander in Afghanistan is General M. I. Sorokin, whose eventual duty is to command the Afghanistan Front when the Soviet leaders decide to advance further into south-west Asia. Sorokin reports directly to General Yu. P. Maksimov, chief of the Southern Theatre of Military Operations, based in Tashkent.

Resistance

Probably 200,000 guerrillas; of this number, possibly 20,000 are intermittently active and perhaps 5,000 are active on any one day. They belong to scores of separate groups, but the main ones are Hizb-i-Islami, which is the most independent and aggressive; the Khalis fundamentalist group; Jamaat-i-Islami; the National Islamic Front of Afghanistan (led by Rahim Wardak, a former army officer); National Liberation Front.

Some outstanding leaders have emerged. Most important are Ahmanshah Massoud, a former engineering student, who leads the guerrillas in the Panjshir area; Ismail Khan, commander in Herat, and Zabihullah around Mazar-i-Sharif. All have large prices on their heads.

Soviet Reporting of the Campaign

The Soviet authorities are highly selective in the details they present to the public at home of their campaign in Afghanistan. Since Soviet forces invaded Afghanistan on 27 December 1979, they are believed to have lost about 10,000 men with a further 20,000 wounded. The Soviet authorities have never published figures of casualties. For the first 2 years of the campaign the Soviet Press revealed little about the duties undertaken by the troops, other than "supply operations" in support of the Afghan forces, and there was virtually no reference to losses. However, with the rotation of personnel and the return home of conscripts and the wounded—and also of the coffins of those killed—news filtered back, and it became impossible for the authorities to maintain complete silence.

On 14 April 1981 the first detailed picture of Soviet military activity in Afghanistan emerged from an article in the Soviet Communist Party organ *Pravda*. Written by a serving officer, it described the dangers of delivering food to remote villages, when convoys were liable to run into mines or be attacked by "counter-revolutionaries". It spoke of the boldness of the soldiers in the face of adversity and the "valuable assistance" they were providing to the local community. No reference was made to Soviet attacks on the Afghan population or to the Soviet Union's political or strategic aims in maintaining its forces in the country.

Subsequent Soviet Press articles described the force as "a limited contingent" carrying out its "internationalist duty". One, in the youth newspaper *Komsomolskaya Pravda* on 7 August 1981, sought to allay families' fears about the young conscripts, taking the form of a response to a mother's letter expressing her concern. The newspaper extolled the soldier's role as a defender of "peace on earth and the happiness and repose of the motherland". The emphasis was again on the army's support for the Afghan people. While this and later articles acknowledged the difficulties under which the Soviet troops were serving, including the rigorous discipline and the extremes of climate, no mention was made of the effects on their mental or physical condition or the injuries and deaths which had been sustained.

On 30 September 1981 the army newspaper *Krasnaya Zvezda* (Red Star) made the first public acknowledgement of a Soviet casualty and articles in *Pravda* and *Literary Gazette*, on 28 October, referred to the active combat role of the troops. A year later there was a further shift in Soviet reporting. Articles in *Krasnaya Zvezda*, on 8 November and 25 and 28 December 1982, admitted that Afghan "guerrillas" had achieved some successes against the Soviet/Karmal regime forces and had inflicted "immense damage". A television film on the third anniversary of the invasion showed a Soviet military convoy under attack and an injured soldier. On 7 January 1984 *Krasnaya Zvezda* described a Mujahideen attack on a supply convoy being escorted by Soviet troops. The following day the newspaper reported an airfield attack on a Soviet aircraft carrying freight. The latter account acknowledged that it was not unusual for aircraft and helicopters to come under fire from the guerrillas and that "each flight mission demands high professional skill and courage. But hardships do not stop the Afghan pilots and transport fliers from the Soviet limited military contingent." The increasing frankness in Soviet reporting on Afghanistan has been accompanied by even tougher language, describing the

Resistance as "terrorists" and "fascists", and by assertions of the inevitability of an ultimate Soviet victory.

Acknowledgements of Soviet loss of life remain rare and are presented carefully, focusing not on numbers killed but on accounts of the exploits of individuals. This was the case in the *Krasnaya Zvezda* accounts of 7 and 8 January 1984. The Soviet authorities evidently expect the struggle in Afghanistan to continue for some time to come. This message is conveyed to the public at home through repeated references in the Press to the Mujahideen as "basmachi"—who were Revolution-period armed guerrilla groups in Soviet Central Asia, across the border from Afghanistan, not finally eliminated until the early 1930s. A recent report on Soviet television spoke of the need to continue "fraternal aid" to Afghanistan while the "revolutionary army" was "still learning how to defend the first gains of the revolution and its fatherland from hostile encroachments from outside".

According to the Afghan People's Support Group, based in London, Soviet troops and the forces of the Karmal regime have killed 19,000 Afghan civilians since the Soviet invasion of December 1979. Pakistan sources put the figure even higher, while American Intelligence reports estimate 30,000 civilians dead.

The Russians do publicise their 170 "aid" projects in Afghanistan, but most of them really benefit the Soviet Union itself. For instance, the Soviet Union's own production of chrome, copper, iron and bauxite is not keeping pace with its needs for home consumption and exports; the remaining reserves are harder to extract and of declining quality. The Russians are therefore exploiting the large reserves in Afghanistan. The copper smelting plant at Ainak is their largest single aid project and all the copper goes to the Soviet Union.

There seems little prospect of an end to the war in Afghanistan. The regime is working on schemes to claim that it has support "of the people" in its desire for the Soviet troops to stay. The government has brought back into use the traditional Afghan *jirgas* (or councils) which have little power but can express opinions. In September 1985 a *High Jirga* of border tribes was held, with 3,700 delegates from frontier tribes attending. This was a large gathering, but no Baluchis were present and only a few representatives from western Pakistan.

Babrak Karmal himself gave the inaugural speech and asked the *High Jirga* to "acknowledge and appreciate the Soviet Union's fraternal assistance". Since the *High Jirga* could not have been held without Soviet approval, it is an indication of yet another Russian move to give legitimacy to the occupation. This, in turn, is likely to lead to further Resistance activity to disprove such a claim. The cycle appears to be endless.

Important Factors for the West

Probably 60,000 Soviet officers have had service in Afghanistan. This should give the entire Soviet war machine a sharper fighting edge than NATO. Nevertheless, a special correspondent of the armed forces newspaper *Red Star* in 1986 criticised the skill and tactics of Soviet troops in Afghanistan. The writer, Colonel V. Filatov, singled out weaknesses in three areas—use of helicopter gunships, artillery support and medical evacuation of the wounded. Throughout his series of articles the military journalist stressed the problems posed by mountains to an army more accustomed to fighting on plains. Helicopters must be able to land on mountain tops; artillery spotters must be trained climbers; medics must know how to get wounded down from mountains.

Late in February 1986 President Babrak Karmal told a British journalist that "To end the war as far as the Afghan people is concerned we need six months—at most one year—to have security in every part of Afghanistan."

"Security" in 1986 was being established at great cost in lives. Atrocities are endless. Kurt Lohbeck, a US television producer who had made six trips into Afghanistan, reported in March 1986 that nearly one million Afghans had been slaughtered. He described how Soviet war planes hammered the Laghman Valley for some days, killing 2,000 villagers. Columns of tanks and commando troops then moved in and killed another 2,000. "Pregnant women were herded together, had their bellies slit open with bayonets and were told to call upon their God to help them. Babies were thrown into bread ovens. Old men were tied down on roads and run over by tanks."

Guerrilla (Civil) War in Angola

LONG CONFLICT IN PROSPECT

Background

The war, which was in progress before independence from Portugal in 1975, became more violent after the Portuguese withdrawal. Foreign involvement added fuel to the conflict. The Popular Liberation Movement of Angola (MPLA) was backed by 13,000 Cuban troops using Soviet equipment; the National Union for the Total Liberation of Angola (UNITA) was supported by South Africa. South Africa was drawn into the war because fighters belonging to the South-West African People's Organisation (SWAPO) used southern Angola as a base for their war against South Africa for the liberation of Namibia.

The two superpowers became involved because of Angola's strategic position. In the event of a major war, the giant oil tankers heading from the Middle East to Europe and the US could be protected (by the US) or attacked (by the Soviet Union) from airfields and ports in Angola.

As American support for UNITA dwindled, the Cubans moved in more troops to help MPLA, which by 1976 nominally held 90% of Angola. Helped by South African advisers, Jonas Savimbi's UNITA troops dominated the more remote regions. While UNITA mainly used classical guerrilla tactics, it sometimes switched to conventional attack. For instance, in August 1983 UNITA besieged the MPLA fortifications of Cangamba for 11 days and then attacked with waves of infantry; 1,300 MPLA troops were killed. Savimbi's tactics were to demoralise MPLA and its foreign allies. The Cubans, 20,000 strong by 1984, were afraid to fight in the jungle.

Savimbi boasted at that time that UNITA was on the point of invading the capital, Luanda. The guerrillas twice cut Luanda's power supply for long periods and South African air and ground attack was supporting UNITA. However, in February 1984 the South Africans and the Angolan government of Jose Eduardo Dos Santos signed the Lusaka Accord which was supposed to end direct South African involvement in the war. There was also to be a phased withdrawal of Cuban troops. South African troops staged a colourful farewell to Angola as they withdrew into neighbouring northern Namibia. Within weeks it was learnt that seventy South African specialist troops had stayed across the border for liaison with UNITA and as a reconnaissance unit.

That the war is still in progress in 1986 is largely because of the military talents and leadership of the lawyer Dr. Jonas Savimbi. A huge, powerful and bearded man, Savimbi speaks fluent Portuguese, French, English and six Angolan languages. A spellbinding orator, he was trained as a guerrilla leader by Chairman Mao's leading instructors and has fought in his country's guerrilla wars for 20 years. He runs a country within a country and, remarkably for a guerrilla force, UNITA possesses established and conventional settlements.

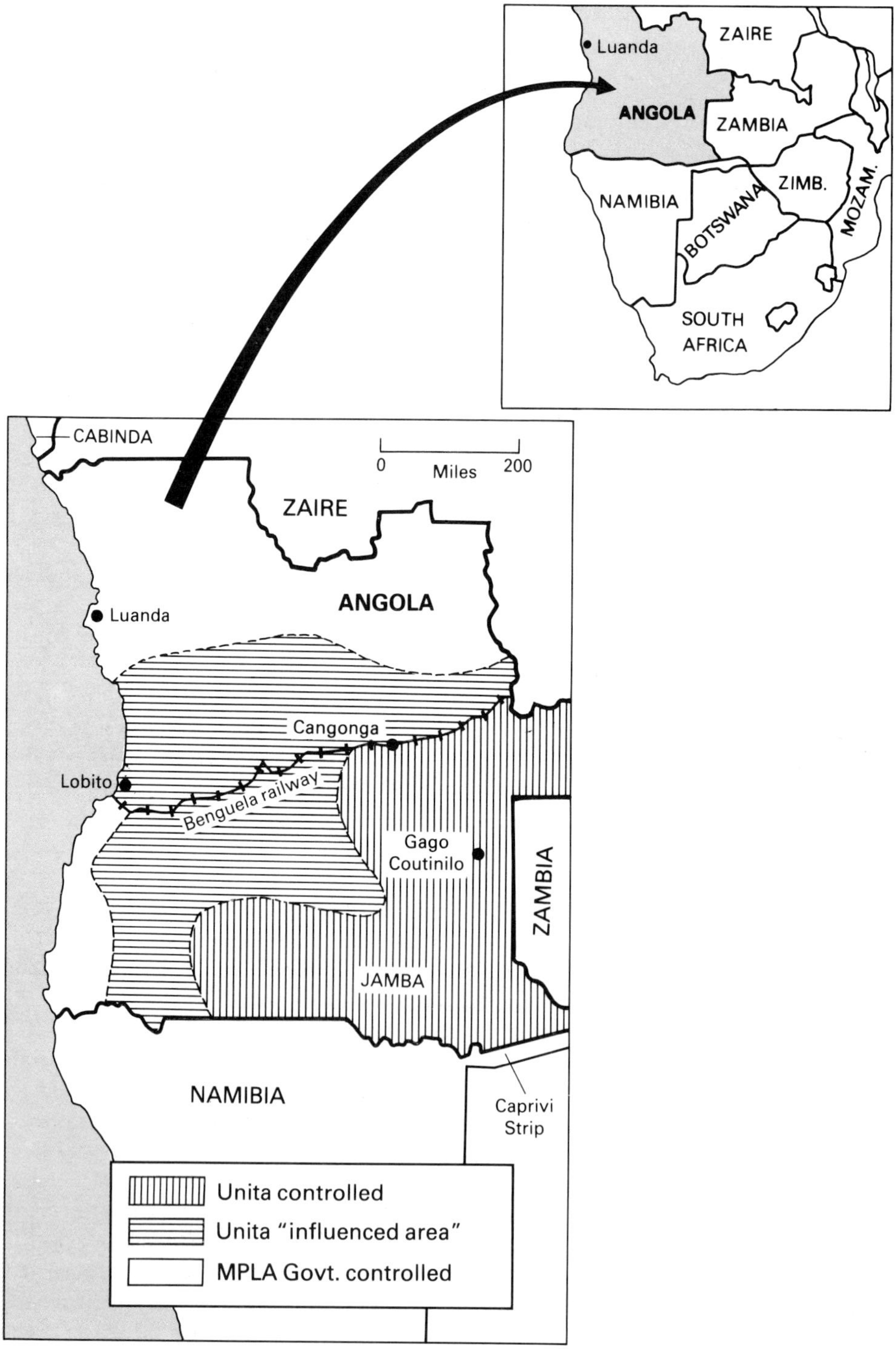

Guerrilla (Civil) War in Angola

The War in 1985

The MPLA's strategy during 1985 was to use more armour and combat aircraft against UNITA, with the intention of breaking the morale of its fighters. The Cubans and Russians, having strongly influenced the reorganisation and training of MPLA, proposed to use more specialist troops from their own forces to back the MPLA assaults. But UNITA could depend on South African help as well as discreet aid from the US and France.

The government forces started 1985 infinitely better trained, equipped and organised than a year earlier. Portuguese counter-insurgency specialists, some of them officers who had served with the Portuguese Army in Angola before it became independent, were brought in to train MPLA units. They created several "Special Brigades", each composed of 500 hand-picked soldiers who were given better food and equipment than the rank and file of the army. They are being used as guerrilla-commandos.

The Angolan regular armed forces total 45,000, including 26,000 conscripts. The army is 40,000-strong and is backed by a militia of 10,500. There is in addition the "Organisation of Popular Defence", which, the government claims, has 500,000 members. These are men who could be called up in an emergency—though they would be without arms and equipment. The army has 165 T-34, 160 T-54 and 132 T-62 main battle tanks, plus 76 PT-76 light tanks. All tanks have a strictly limited use in much of jungle-clad Angola.

The air force commander, Iko Carreira, a Portuguese-trained pilot, has thoroughly restructured his force, which has 90 combat aircraft, mostly Mi-21s, Mi-23s and Mi-25s. Cubans and Russians have trained the Angolan pilots in ground-attack techniques similar to those used by the Russians in Afghanistan. Key air defences in southern Angola have been greatly improved with Soviet-supplied equipment. The Lubango basin is best defended of all, with so many SAM missiles that South African aircraft avoid the region. (In 1983 the Angolans claimed that their air defences had brought down seven South African aircraft.) The nerve-centre for the defence of southern Angola is Lubango airport, where many Cubans and some East Germans are on duty. Aircraft activity is intense, with Mi-23 fighters frequently flying in or out. Soviet Antonov transports by the *score* arrive almost daily, carrying either arms, troops or supplies.

Apart from the continuing supply of arms and equipment, it is possible to see the beginning of a decline in Angolan dependence on Cuba and the Soviet Union. The younger more professional Angolan officers, especially those in the air force, certainly want to be more independent; this is an indication of their self-confidence. About 80% of the combat pilots are Angolan, compared with 40% in 1982.

However, too much should not be read into this spirit of independence. There are still 25,000 Cuban troops providing "defensive support" along the southern border with Namibia. Generally, the Cubans are careful to avoid clashes with South African and UNITA troops, though the pilots are sometimes in action. At least 5,000 Soviet servicemen are on duty in Angola and tank crews in particular have been in action.

The President of Cuba, Dr. Fidel Castro, has told the UN Secretary-General that Cuban troops will remain in Angola until UN Security Council Resolution 435, leading to independence for Namibia, is implemented. Cuba backs and supplies SWAPO.

In May 1985 the South African Army sent three commandos into Angola on a mission "to gather intelligence about African National Congress and SWAPO training camps". As the commandos were equipped with contact and incendiary mines, firearms with silencers and shortwave radios the stated mission was obviously a cover. The Angolans ambushed the patrol, captured one soldier alive and were able to prove that the men had been sent to sabotage Gulf Oil installations in the Cabinda enclave. The commandos were found with UNITA leaflets, and had their mission been successful the impression would have been that it was the work of Savimbi's guerrillas. Other attacks on the state-owned oil refinery in Luanda and on power lines outside the capital have probably been the work of South African commandos.

In July 1985 South African troops killed 45 SWAPO fighters during a "hot pursuit raid" into Angola. The SWAPO guerrillas had attacked a South African base at Eenhana, east of Oshikango, on the Angolan-Namibia border.

The real teeth of the South African campaign in Namibia and Angola is the little publicised 32nd Commando Battalion, formed mostly of Portuguese-speaking blacks. The unit, 2,000-strong, is based in the Caprivi strip, a finger of Namibian territory between Angola, Zambia and Botswana. Fanatically opposed to the Luandan regime, these troops have been welded into an élite force which has a large proportion of foreign professional soldiers among its officers and NCOs. Because of its non-South African make-up, the 32nd can be committed in Angola without alarming South Africa's white population over the loss of young conscripts in battle; the anti-conscription lobby is strong in South Africa.

Savimbi is crucial to South Africa's security interests. His control over the south-eastern section of Angola effectively seals half of the long border with Namibia to infiltration by SWAPO guerrillas. It also ties down 5,000 SWAPO men who help to defend Angolan bridges, railways and power lines against UNITA saboteurs in return for being allowed to conduct their Namibian war of independence out of Angola. If the UNITA threat to Angola were reduced or eliminated, South Africa would face a greatly increased guerrilla war in northern Namibia.

In August 1985 the government launched an offensive—its heaviest ever against UNITA—to capture Jamba, Savimbi's main base. A sprawling, well-developed bush settlement about 100 miles north of the Namibian border, Jamba has vehicle repair debots, a hospital, a clothing factory and accommodation in huts for 10,000 people.

Soviet officers planned and commanded the tactical operations and Russian crews manned some of the tanks. The 8th, 13th, 14th and 21st Brigades were the teeth of the campaign. In the 45-day campaign the MPLA made a two-pronged thrust, using Soviet tanks, with air support from Mi-24 helicopter gunships. A Luandan communiqué claimed that 1,300 UNITA fighters were killed. However, Savimbi's men, together with the South African 32nd Battalion, beat off attempts to capture the towns of Moxico and Mavinga in Cuando Cubango province. Savimbi had a small triumph when his troops shot down a Mi-21 Fishbed aircraft of the Angolan Air Force.

While MPLA did not capture Jamba, the offensive shifted the balance of power towards the government. The sustained attack did not break UNITA's morale, as the government had hoped, but it was a reverse for Savimbi. Loss of Jamba would be a heavy though not necessarily a fatal blow, and he now knows that it is vulnerable.

Faced with this threat to its ally, South Africa's support for UNITA at once became less covert. After years of denial, the Defence Minister, Magnus Malan, admitted in September that South Africa did give aid to Savimbi and warned that it was prepared to intervene militarily and "without limit" to save him. Military analysts in Johannesburg believe that the September 1985 offensive reflects the tougher attitude in Moscow since Mikhail Gorbachev became the Soviet leader.

The Angolan government claims that only the seasonal rains saved Jamba by bogging down the mechanised advance. UNITA leaders expect a renewed campaign after April 1986. It can be no coincidence that after a 9-year ban on aid to UNITA the American government started openly helping Savimbi's force in September 1985. This led to the Angolan government breaking all diplomatic links with Washington; it said that American aid to the rebels and South African "belligerence" prevents Angola from cutting its links with Soviet bloc countries. However, the Soviet bloc would not withdraw its troops even if asked. The whole situation in southern Africa is so volatile that the Soviet Union is poised to take advantage of any crisis—such as the security problems which could threaten South Africa and which do threaten Mozambique.

Short of killing Jonas Savimbi, the Angolans and their allies cannot defeat UNITA. He has shown himself to be a brilliant guerrilla leader, probably the most efficient the African continent has ever produced. His highly experienced subordinate leaders are loyal and enterprising and the fighters are tough and strongly motivated. UNITA is almost impossible for MPLA agents to infiltrate; at least thirty who tried to do so were quickly exposed and executed. Angola covers an area of 1,247,000 km^2, much of it jungle-clad, so that any guerrilla force has much freedom of movement. The Angolan civil war could last for many more years.

When Savimbi publicly revealed his strategy in February 1986 he clearly had a long war in mind. "Our strategy is to raise the costs of the occupation of Angola until the Cubans and Russians can no longer bear the burden. A central element of this strategy is to deny the colonial forces the revenues that finance their occupation. The Russians and Cubans exploit Angola with a rapacity unrivalled by the Portuguese. Soviet fishing fleets have swept out coastlines and fish has disappeared from the Angolan diet. Our diamonds, minerals and oil are taken in 'payment' for arms and support of the Cuban troops, a total of $480 million a year. These Cuban mercenaries can afford to stay in our nation only as long as they are paid. So UNITA attacks the diamond mines, sabotages the bridges and destroys the industries that support the Cuban occupation. . . . When enough Cuban officers are returned to Cuba in coffins then Castro must face the wrath of his own people." At the same time Savimbi not only admitted to receiving South African help but support from China, Arab nations and black African countries.

In March 1986 a new Cuban general, General Ramirez, arrived in Cuando Cubango province to command government offensive against UNITA. The Angolan government had been strengthened by MiG-23 and Sukhoi Su-22 aircraft; they could make it more difficult for UNITA to repulse Cuban-Soviet-MPLA offensives without more American aid.

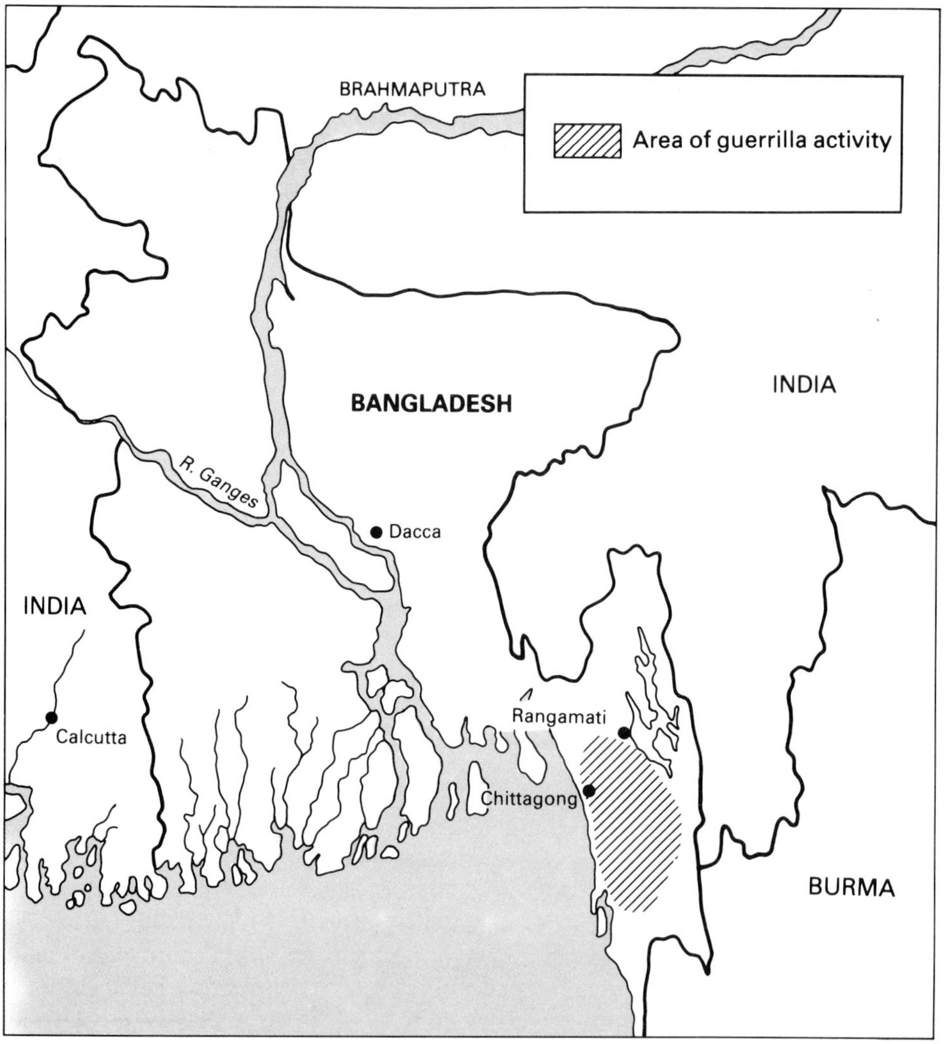

Bangladesh Guerrilla War

Bangladesh Guerrilla War

CONFLICT IN THE HILL TRACTS

Background

For half a century before the partition of India and Pakistan in 1948, the British ruled the Chittagong Hill Tracts in what nearly everybody considers to have been an enlightened way. The inhabitants of the area were nearly all Buddhists, from twelve different tribes, with Chakma people predominating. The British planners made a serious mistake when deciding on the border between the two newly independent nations, India and Pakistan. The Hill Tracts and their Buddhist people were handed over to Pakistan, a martial, theocratic society, rather than to democratic, secular India.

In 1951 there were ten Buddhists to every Bengali Muslim. At first there was no demographic problem, but with the founding of Bangladesh in 1971 the Chakmas, a people without a martial tradition, came under systematic attack when the new government decided to settle the Hill Tracts with Muslim/Bengalis from the crowded river plains of the Ganges and Brahmaputra. More than 200,000 Bengalis were settled in Chakma country. Several responsible organisations, such as the Anti-Slavery Society and Survival International, have alleged large-scale massacre of Chakmas. Being Buddhists, the Chakmas are a gentle people and are Mongoloid in appearance; these characteristics are resented by the East Bengalis, who consider themselves racially pure.

In 1974 a Chakma self-defence force was formed—the Shanti Bahini guerrillas. It was armed by the Indians and by the Russians. It quickly became a ruthless fighting force, raiding Bangladeshi banks, Bengali settlements and army posts and ambushing Bangladeshi units on jungle trails. In January 1984 the Shanti Bahini kidnapped three Shell oil company surveyors from their jungle camp and kept them until paid a ransom of $1 million.

The army has sent division after division into the jungles to wipe out the Shanti Bahini and in 1984 posted the crack 24th Division, the "Bengal Tigers", to the Hill Tracts. Major-General Noor Uddin Khan was given *carte blanche* to "solve the Hill Tracts problem".

The War in 1985

By 1985 the war was even more bloody and vicious. Survival International estimated that by now 185,000 people had died in a campaign of genocide. Bengali/Muslims now made up 45% of the population.

In the continual fighting the army was losing a soldier every week and claiming that it was killing two tribesmen a day. Scores were being hurt and many had their homes and property destroyed or stolen; the government's objective, if it cannot kill all the Chakmas, is to drive them out of the region. Survival International claims that up to 15,000 people have been detained without trial and that many have been tortured.

To compound the confusion there are dissident Indian groups in the Tracts; arms reach them from the Chinese and the Americans. The Americans help the Indians because they fight the Soviet-supplied Shanti Bahini.

During the year the Shanti Bahini guerrillas, who number perhaps 6,000, found themselves fighting not only the 24th Division but the police and paramilitary units, perhaps 125,000 men in all. The ground forces are helped by helicopters of the Bangladeshi Air Force, but the aid they can give in the mountains and rain forests is limited. This is not an area where gunships can inflict casualties.

The Shanti Bahini guerrillas, despite the odds against them, are well led and formidable jungle fighters. Even the "Bengal Tigers" fear guerrilla ambushes, which are so efficient that often a Bangladeshi patrol will lose ten men without ever setting eyes on a guerrilla. The Shanti Bahini fighters are also highly proficient with booby traps; as night raiders, to cut throats or steal weapons and supplies, they are unsurpassed. Unable to defeat the guerrillas, the army wreaks vengeance on the villagers who supply them with shelter, food and information.

Few reporters have ventured into the Chittagong Hill Tracts, but those who have done so believe that the war will continue for another 10 years—or until the Bangladeshi government achieves what some people see as its policy of genocide.

Burkina Faso–Mali

THE CHRISTMAS WAR

On Christmas Day 1985 the armed forces of Burkina Faso (formerly Upper Volta) invaded northern Mali as the climax to a frontier dispute which began in 1965. The leader of Burkina Faso, Captain Thomas Sankara, was seeking to capture five frontier villages held by Mali.

Burkina aircraft bombed Sikasso, a small town 25 miles inside Mali, where the Malians have a military base. Burkina's tiny air force consisted mostly of old French and British transport planes and nobody had realised until the raid that it had received some Soviet MiGs. Sankara has run Burkina on Marxist lines since he took over in a coup in 1983.

Provocation has come from both sides of the border, with the Algerians supporting Mali and Colonel Gaddafi of Libya backing Burkina Faso. It was soon clear that Gaddafi was using the Christmas war to extend his influence in West Africa and to further destabilise the more conservative countries in the area. Gaddafi blames the "French colonial heritage" for the conflict and warned France to stay out of the dispute.

On Gaddafi's insistence, Captain Sankara refused to attend the summit of Francophone African countries in Paris earlier in December. Instead he played host to Gaddafi who offered to "fuse together" Burkina and Libya. For impoverished, starving Burkina the offer is attractive.

Gaddafi also visited Bamako, the capital of Mali. The Malian government claims that it was while Gaddafi was in Mali on a so-called goodwill mission that his protegé Sankara moved his troops across the border. Gaddafi's visit, the Malians say, was nothing more than a tactical ruse to lull Mali's defenders into a false sense of security.

Mali, which is supplied with French arms and equipment and has French military advisers, is the stronger nation of the two, but now that Burkina has MiGs the situation is more dangerous.

Both the landlocked countries are deeply in debt; 80% of Mali's gross national product goes to pay the interest on the nation's debts. Burkina and Mali fall within the Sahel zone where long drought has dried up the land; thousands have died of starvation.

The war lasted for 3 days and the fighting, mostly with small-arms, was savage. At times the combatants fought hand-to-hand in the streets of the contested villages. An estimated 100 soldiers and 300 civilians were killed before a cease-fire, arranged by Algeria, came into effect on 28 December. Mali and Burkina agreed to refer their dispute to the international court in The Hague which is expected to announce its verdict in June.

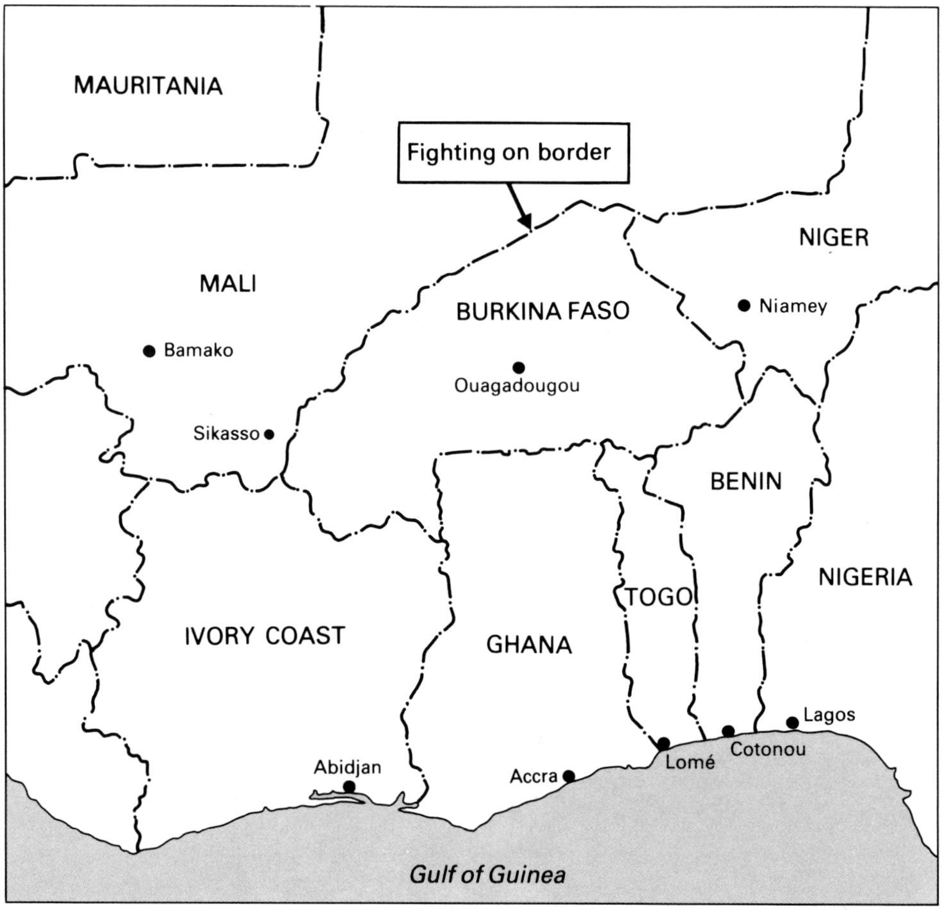

Burkina-Faso/Mali War

Burma Guerrilla War

"KINDNESS AND KILL CAMPAIGN"

Background

Since 1949 the Burmese Army has been waging a counter-insurgency war against the Karen people. It is one of the longest single guerrilla campaigns in history, and there is no sign of its ending. The 4 million Karens demand independence for their region which they call the "Republic of Kawthoolei". Burma is a one-party, socialist non-aligned state; the government does not recognise the rights of a minority to be politically represented in government.

The Karen National Liberation Army (KNLA) is only one of six groups fighting the government for complete autonomy for their regions. The Kachins, Shans, Arakanese, Mons and Karreni tribes are loosely grouped with the Karens in the National Democratic Front (NDF). The NDF has been in existence since 1976, but has not managed to create a unified military command.

The main battlefield is the Irrawaddy Delta where the KNLA controls a large area along the border with Thailand. About 4,000-strong, the KNLA divides its fighting men into five "brigades". They finance their war against the Rangoon government by imposing a 5% toll on traders passing through customs gates from Thailand with blackmarket goods, mostly opium.

To oppose these highly experienced and well-led guerrillas, the Union of Burma Army has 170,000 men, comprising 89 battalions of the Burma Regiment, 6 battalions of the Burma Rifles and 17 other rifle battalions. The Light Infantry Regiment has 9 battalions. Their only modern equipment are small arms—Kalashnikovs, M16s, Armalites and Uzis. The army has old British-built Comet tanks, Humber and Ferret vehicles, 76-mm Yugoslav mountain guns, 105-mm American howitzers and anti-tank guns. None of this equipment is much use against the Karens who have operated for nearly two generations in the classic guerrilla fashion with hit-and-run attacks on army bases and communications.

The War in 1985

During 1985 the army, in a new attempt to end the 38-year war, evolved a two-part strategy—what is unofficially called "the kindness and kill campaign". The kindness aspect is a drive to win the "hearts and minds" of the Karen villagers; the kill strategy is to crush one ethnic group at a time rather than dissipate the army's resources by attacking all the factions simultaneously.

The hearts-and-minds campaign is supposed to be modelled on British tactics in Malaya during the emergency of the 1940s, but it is not having the same success. Detachments tour all the villages in a "target" district and officers give lectures about the merits of the Burmese Socialist Party and the evils of the Karen

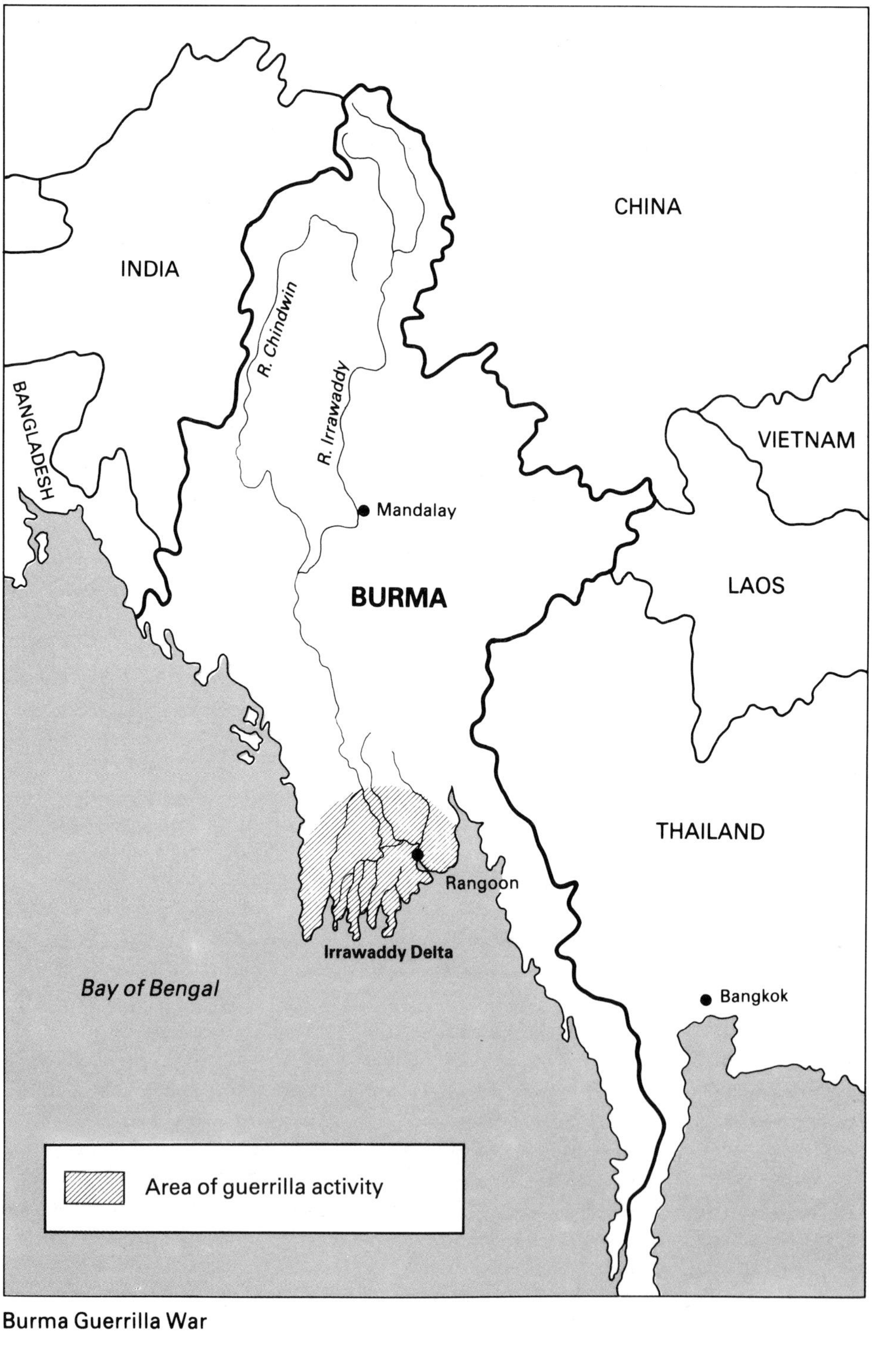

Burma Guerrilla War

leadership. The villagers are urged not to support the guerrillas. The highlight of each of those visits is supposed to be a meeting between the "Column Commander" and the village chiefs, but few of them bother to turn up.

These public relations operations are dangerous, so strong escorts accompany the goodwill teams. Unfortunately for the ideals of the programme, the troops' behaviour is counter-productive. They often steal pigs, chickens, rice and money. Sometimes a commander will confiscate an entire crop. On one occasion, while an officer was trying to win over the villagers with bland words, another officer was executing a man found with a rebel symbol tattooed on his arm.

Killing Karen guerrillas is difficult. Their leader, General Bo Mya—who is also president of the National Democratic Front alliance—is an able military commander. Indeed, though little known, he is one of the world's most successful guerrilla leaders.

General Bo Mya has taught his men not to be over-confident, and he has patrol officers whose entire professional role is to ensure that the guerrillas obey the rules of classic irregular warfare. For instance, Karen fighters never stay two nights in the same place—they do not even camp a second time in a spot once used. To sleep, they spread out in twos so that no grenade or shell can wipe out an entire unit; they never light fires after dark, and they move along shallow rivers so as to leave no trail. The patrol officers make unannounced "raids" to check the men's vigilance. During 1985 Mya was himself confident enough to establish a jungle and river warfare training camp—which the Burmese army cannot find.

Some raids by the KNLA are fairly large scale. The 7th Brigade attacked government buildings, the airport and the military base in the town of Papun. They caused much damage and inflicted 100 casualties without loss to themselves. After 38 years' active service, the KNLA men know every metre of the difficult terrain. Army officers have nothing better than old British Army or Japanese Army maps dating back to the jungle war of 1941–5.

The government has withdrawn administrators from some large regions, a tacit admission that the situation is out of control.

Another indication of this is that in 1984 much of the conflict was taking place close to the Karen strongholds. During 1985 the action moved deeper into Burma as the Karens went on the offensive, forcing the army units to leave the Karen region and chase the guerrillas.

The success of the KNLA has not gone entirely unnoticed. The Contras of Nicaragua and at least one unidentified faction from Angola have sent agents to the KNLA headquarters in Manerplaw to ask Mya if they may study his methods. His answer was a polite but firm rejection. Mya's success is largely the result of good security; he does not discuss his methods.

Nevertheless, Mya has made it clear that he might be willing to trade training for modern heavy weapons to replace the antiquated and sometimes village-made weapons such as mortars. Meanwhile, the Rangoon government and Burmese army is seeking professional help, mainly from China, in its desperate attempts to defeat the Karens and end the war.

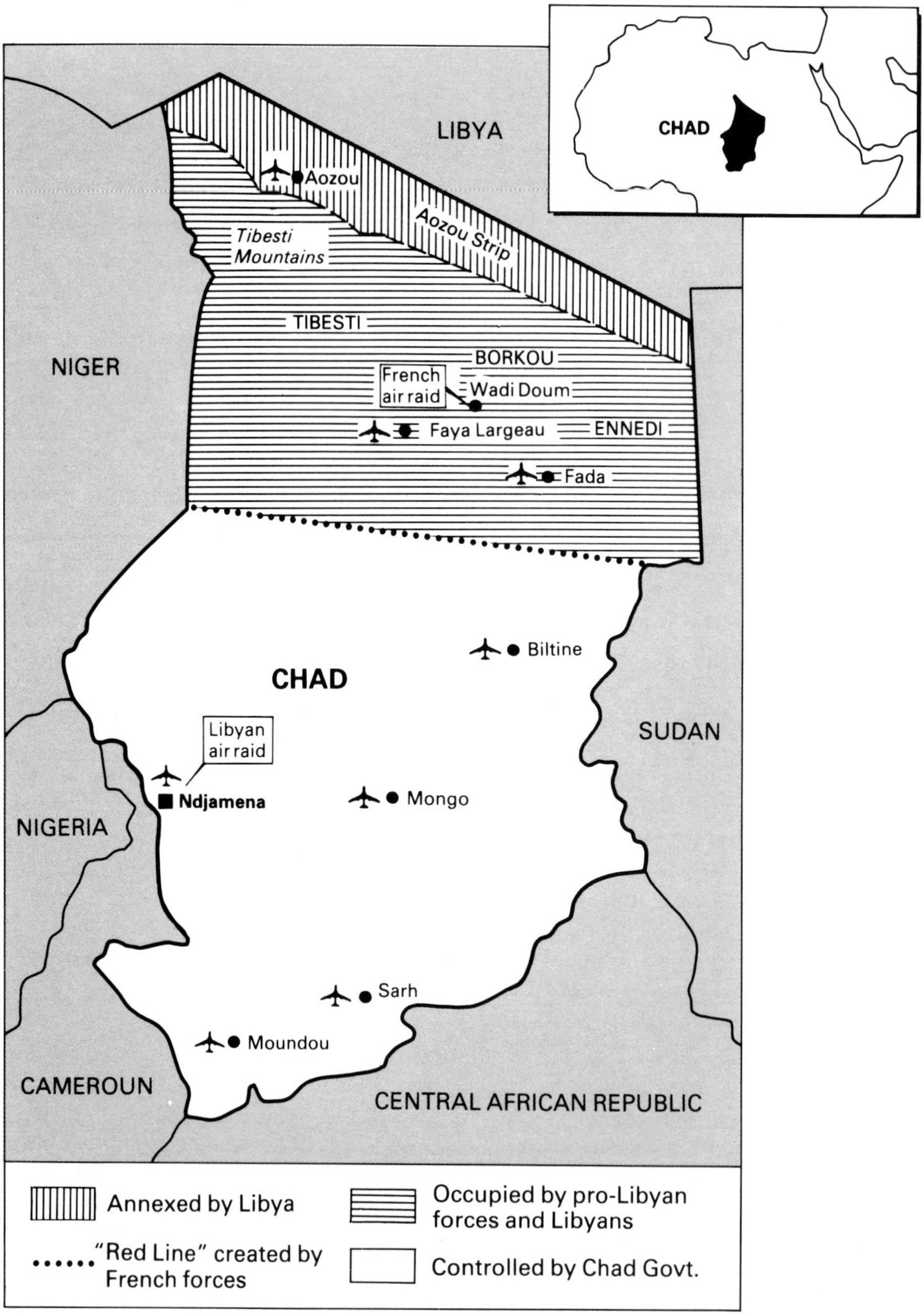
LIBYA
CHAD
Aozou
Aozou Strip
Tibesti
Mountains
TIBESTI
BORKOU
NIGER
French
air raid
Wadi Doum
Faya Largeau
ENNEDI
Fada
Biltine
CHAD
SUDAN
Libyan
air raid
Ndjamena
Mongo
NIGERIA
Sarh
Moundou
CAMEROUN
CENTRAL AFRICAN REPUBLIC
Annexed by Libya
Occupied by pro-Libyan
forces and Libyans
"Red Line" created by
French forces
Controlled by Chad Govt.

Chad

War in Chad

THE 25-YEAR WAR

Background

A landlocked and mostly arid country of 700,000 km^2, Chad borders six states: Libya to the north, Niger, Nigeria, Cameroun to the west, Central African Republic to the south and Sudan to the east. This central position is Chad's strategic value and it is largely the reason why it interests outsiders.

The population of 5 million comprises numerous ethnic groups, with Muslims predominant in the north and Christians in the south; the Christians are in the majority. Formerly part of French Equatorial Africa, Chad became independent in 1960, but the three northern-most provinces remained under French military administration until 1965.

After much post-independence conflict, which began in 1960, the various political groups met in Khartoum, August 1978, to establish a broadly based central government for Chad. This and subsequent efforts failed and the factional war went on between the Forces Armées Populaires (FAP) of Goukouni Oueddei and the Forces Armées du Nord (FAN) of Hisseine Habré.

Within months of Colonel Gaddafi's coup against the monarchy in Libya in 1969, he began intervening in Chad. In 1973 Libya occupied the Aozou Strip, a 60-mile wide stretch of mountainous desert in the north. Gaddafi has frustrated attempts to produce a stable and independent government in Chad, supporting Habré against Goukouni, Goukouni against Habré and fomenting trouble in the Christian south against Habré. In mid-November 1980 a Libyan tank force entered Chad and a Libyan air base was soon under construction 60 km north of Ndjamena, the capital. At least 15,000 Libyan troops were involved in fighting against Habré's FAN. FAN regrouped in the north-east and with help from Egypt, Sudan and the US took the offensive. On 6 January 1981 a joint communiqué issued by Libya and Goukouni's Gouvernement d'Union Nationale de Transition (GUNT) announced that Libya and Chad would work in collaboration for "people's unity".

The Organisation of African Unity (OAU) denounced the proposed merger, demanded the withdrawal of Libyan troops, called for free Chad elections and sent a peacekeeping force with financial and material aid from Western countries, including Britain and France.

Habré's forces captured Ndjamena in June 1982 and Goukouni fled into exile. That month, as Habré formed a government and began to revive public services, banking, commerce and education, the OAU force withdrew.

Libyan aggression continued into 1983, in support of Goukouni. Gaddafi demanded a declaration of Chad's "Arabness"—though Arabs are in a minority—close integration with Libya and recognition of Libyan possession of the Aozou Strip.

In May 1983 GUNT forces, with Libyan help, attacked and captured the southern city of Faya-Largeau, then Oum-Chalouba and Abéché. Zaire sent paratroopers and fighter aircraft to aid Habré; the US supplied $25 million worth of arms and equipment. The Soviet Union declared unqualified support for GUNT and Libya.

Responding to appeals from Habré, France sent troops to Chad under an existing defence agreement and on 13 August 1983 established a "Red Line" at the 16th parallel. The Libyans, in retaliation, shot down a French aircraft.

Following tortuous diplomacy and many quarrels, Gaddafi announced on 30 April 1984 that he would withdraw the Libyan troops "if that had been the pretext for the French presence". In August the French stated that their troops would not leave Chad as long as any Libyan soldiers remained. The Zaire units were still in Chad, apparently without any objections.

The most acceptable head of state within Chad at the end of 1984 was Habré; only he seemed able to produce a stable, broadly-based government supported by a majority of Chadians and by the OAU.

In Cyprus, in September, President Mitterrand and Gaddafi signed an agreement that both their countries would withdraw their troops. The French kept to the treaty; the Libyans did not.

The War in 1985

By January 1985, with the war 25 years old, Goukouni was back in Chad with his headquarters at Faya-Largeau (but now called Faya-Abouchar), an oasis 60 miles long. He made it clear that with Libyan backing he would destroy Habré, whose army was 150 miles away to the south.

In June Libyan workers hard-surfaced the immensely long runways of a new airfield in the north. Libya had been able to reach the south of Chad only with long-range Soviet aircraft; the new airstrip gives shorter-range planes the ability to fly farther south than ever before.

In southern Chad Libya dropped arms to aid commando units which had mutinied—with Libyan incitement—against Habré. France responded by giving Habré $500,000 to pay allowances to any of Goukouni's commandos who might want to abandon him and join the government forces. In the fiscal year to 1 October 1984, the US gave Habré's Chad $52 million to economic aid, but made it clear that France has the primary role in this part of Africa.

In August 1985 President Sasso N'Guesso of Congo invited the various Chadian leaders to a conference in his capital Brazzaville. He was acting at the request of the OAU which was trying yet again to end the interminable civil war. The meeting was abortive, because Habré did not turn up. He had instead travelled from his capital Ndjamena to Abéché, scene of several battles, to review two commando battalions trained in Sudan and Zaire. In a speech he warned of the "Libyan threat".

According to Western satellite intelligence, Libyan troops are still operating with Goukouni's men—in an area they should have evacuated in line with the Mitterrand–Gaddafi agreement. The Libyans are members of Gaddafi's "Islamic Legion", the force which is to move on from Chad in due course and conquer other states.

The French have not again sent troops to Chad, though Habré has French military advisers. However, French units are stationed in the neighbouring Central African Republic and could quickly reach southern Chad.

At the end of 1985 Habré was holding the Libyans in check in the north, but he had troubles in the south where several rebel groups oppose him. The most powerful is Codos Rouges, commanded by Colonel Alphonse Kotiga, who held government posts in the 1970s. Kotiga, who sees himself as president of Chad, led an attack on Cotonchad, which handles Chad's main export, cotton. In 1984 Chad made $11.5 million from its cotton exports—second only to Egypt in Africa—but in 1985 this amount had fallen to $4 million because of civil war disruption.

Kotiga will not negotiate with Habré, so there are now three main armies operating in Chad, as well as the Libyans. The French, and others, might well be forced to intervene to end the war which has torn apart one of the poorest countries in the world.

On 16 February 1986 the French made an air strike which destroyed the Libyan-built airfield near Wadi Doum in the north. The following day a Libyan Tupelov-22 jet bombed the airport at Ndjamena. Neither raid seriously affected the opponent's ability to make war. Instead, their purpose was to reassure allies and warn the opponent. Nevertheless, in the wake of the Libyan attack France reinforced its forces at Ndjamena in an operation code-named Operation Epervier (Sparrowhawk). Six Mirage F-1 fighters, four Jaguar strike aircraft and 750 personnel make up the garrison at the Chadian capital. The message to Gaddafi was clear—Do not move further south.

Gaddafi's counter-attack showed that his planes could still strike southern Chad and that he would not abandon his GUNT allies.

Despite a Libyan–French understanding that the 16th parallel would not be crossed by Libyan-backed Chadians, they twice did so in February 1986 and "battles" took place at Oum Chalouba and Kouba Olanga. Habré's troops had the better of the encounters. Should they start losing against the Libyan–GUNT coalition there was little doubt that France would send in its crack rapid deployment force based in the Central African Republic. Far from falling back northwards, the GUNT–Libyan troops, in three widely separated locations, were often reinforced and strengthened by heavy equipment. By mid-1986 Gaddafi had still not moved into the offensive phase which the French expected.

Colombia Guerrilla War

Colombia Guerrilla War

A GENERATION OF CONFLICT

Background

A civil war between the Conservatives and the Liberals began in 1948. This conflict drew into it the Fuerzas Armadas Revolucionarias Colombianes (FARC), or Colombian Revolutionary Armed Forces. FARC is the military arm of the Colombian Communist Party.

Under the influence of Castroism, two other guerrilla groups came into being in the 1960s: the pro-Cuban Army of National Liberation (ELN) and the Maoist People's Liberation Army (EPL). In 1973, as the fighting continued, the M-19 movement was founded; a leftist-nationalist-Castroite group, it quickly became powerful. In 1982 yet another guerrilla group was set up—the Trotskyite Autodefensa Obrera, or Workers' Self-Defence, but better known as ADO. The five groups had just one thing in common—the end of ordered government, the growth of anarchy and the development of a new social system.

In 1982 Belisario Betancur, a Conservative, became president and pledged himself to secure peace. In 1984 a ceasefire was signed between a government-appointed peace commission and all the groups except ELN. A majority of dissidents in all the other groups did not respect the ceasefire, despite its generous terms. It was for a 1-year trial period during which Betancur hoped for a "great national dialogue". The guerrillas were permitted to keep their organisational structures, their arms and their uniforms.

Betancur was trying to follow the example of Venezuela, whose government had succeeded during the 1960s in pacifying and domesticating its guerrillas by way of an amnesty for the surviving remnants of guerrilla groups which had already been militarily defeated. The Colombian case is seriously different—here the guerrillas have never been decisively beaten. Betancur was working on the assumption that the guerrillas are genuinely interested in social reform rather than simply in gaining power.

The War in 1985

An M-19 spokesman declared early in the year that the ceasefire is "neither the dissolution nor the strengthening of the guerrilla groups but simply the maintenance of the status quo at the moment the agreement was signed".

This was an indication that M-19 and the other groups had drawn the inference that the president's desire for peace was a sign of weakness. As a result, they made many demands, such as the replacement of certain government ministers and the withdrawal of troops from rebel-held areas. When Betancur refused, both FARC and M-19 threatened to launch a "popular rising".

In March 1985 M-19, EPL and ADO announced that they were "temporarily" withdrawing from the dialogue. The hard-working and sincere Betancur, having already committed the government to vast social and technical infrastructure projects in the guerrillas' areas, was frustrated. The guerrillas constantly try to make the president look incompetent, ineffectual, powerless and foolish. During a state visit to Mexico, Betancur held talks with Ivan Marine Ospina, who was then head of M-19. Almost immediately after this meeting, in a statement broadcast on Mexican television, the guerrilla commander encouraged Colombian narcotics dealers to make good their threat to murder US diplomats. He succeeded in giving the impression that this was the desire of the president.

The president is also under attack by his own colleagues and the army chiefs for not disarming the guerrillas. In the last year 207 soldiers and police were killed in clashes with guerrillas and 900 were wounded.

Despite the ceasefire, rebel attacks continued throughout 1985. Bombings, robberies, murders and kidnappings for ransom money are commonplace. During protest demonstrations by EPL and ADO much violence has taken place, and these two groups are also notorious for abductions.

Elements of FARC and M-19 had pitched battles with army units in three separate provinces in April 1985, and later government troops laid siege for 3 weeks to a major M-19 stronghold near Carinto, Cauca province. After many casualties on both sides, the guerrillas slipped through the army's lines by night.

In the capital, Bogota, M-19 is considered to be the worst violator of the ceasefire. The group's M-19 supreme commander, Alvaro Fayad Delgado, says that the government's insistence that the guerrillas lay down their arms is "humiliating" and threatens reprisals. M-19 is heavily supplied by Cuba and works closely with Colombia's narcotics underworld which provides money for the guerrillas' arms purchases. The country's cocaine king, Carols Lehder, son of a German and a pro-Hilter propagandist, recently publicly declared that he has made "common cause" with M-19.

Many analysts, inside as well as outside Colombia, believe that President Betancur made a fundamental mistake in signing the peace treaty with the guerrillas. This, his critics say, politically recognises the men of violence. The treaty restricts the activities of the government and security forces while granting the terrorists more freedom of action.

The Colombian Army has 57,000 men, a navy of 8,500 and an air force of 4,500. In addition, there is a para-military national police force of 50,000. Personnel are well armed and the army has 200 EE-9 Cascavel armoured cars and 45 armoured personnel carriers. The air force has fewer than 40 helicopter gunships, but during 1986 will get 10 more. These are the only aircraft suitable for anti-guerrilla warfare in Colombia's 650,000 km^2 of varied terrain.

Unlike guerrillas in several Central American countries, the Colombian fighters do not have popular support either in urban areas or the countryside, but the groups' punishments for "failure to co-operate" ensure that they get enough assistance, no matter how unwillingly given. Hundreds of farms have been abandoned by landowners forced to pay "taxes" to the rebels.

FARC has been fighting for a generation, and its veterans are tough, experienced and skilful. This also applies to the army, which does not need advice from foreign instructors.

Terrorism intensified throughout 1985, partly because parliamentary and presidential elections were due in May 1986. As the combined Colombian left makes up no more than 15% of the country's voters, extremists feel that the only way to power is through a "war of liberation".

The government's knowledge of some of the groups was increased by the defection of Luis Alberto Rodriguez, deputy commander of FARC's "Eleventh Front". He revealed that FARC has organised a 36,000-men army made up of 600 "fronts", each with strength of 600. These are to be the striking units. The figure alarmed the government, as it had been generally accepted that overall the number of guerrillas was no more than 16,000. Independent intelligence indicates that 36,000 for the new FARC force is an exaggeration, but the army's strategy has had to change to take account of much greater numbers of rebels.

The armed conflict—the longest-running guerrilla war in Latin America—has not touched the great majority of Colombia's 29 million people. Nevertheless, it has resulted in 20,000 deaths and in scores of thousands being injured. Military psychologists in Bogota believe that the war is unlikely to end because it is a way of life—and a livelihood—for so many guerrillas. Only the complete withdrawal of Cuban and Soviet-bloc military aid can end the war.

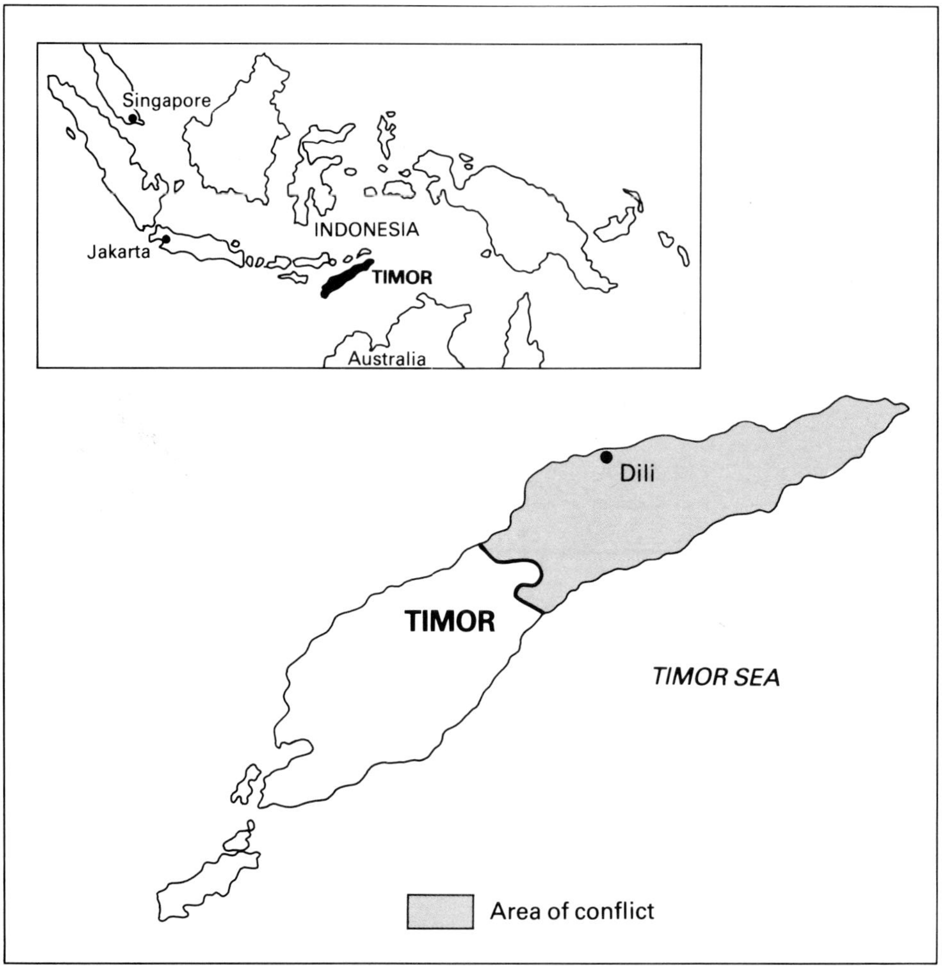

East Timor Resistance War Against Indonesia

East Timor Resistance War Against Indonesia

"COMMITTED TO GENOCIDE"

Background

In 1975 the Portuguese, having possessed "Portuguese Timor" for 500 years, abandoned the place. They already had all the trouble they could handle with wars in Angola and Mozambique and demands for independence from their smaller colonies.

The Portuguese had shared the eastern end of the island of Timor with the Indonesians, an area known as Timor-Dili. The western half had been Dutch and in common with all other Dutch colonial possessions in the East Indies had become Indonesian territory.

Various factions sought to control the former colony of 600,000 people, inhabiting an area of 15,000 km^2. The most powerful group was the Marxist-orientated Frente Revolucionara de Timor-Leste Independente (Fretilin); the others were the arch-conservative Unaio Democratica de Timor (UDT) and the minority Associacao Popular Democratica de Timor (Apodeli) which favoured integration with Indonesia.

In a despatch dated July 1975, the British ambassador in Jakarta, Indonesia, Sir John Ford, recommended: "It is in Britain's interest that Indonesia should absorb the territory and as unobtrusively as possible, and if it should come to the crunch and there is a row at the United Nations we should keep our heads down". The report was later published in Australia and embarrassed Britain.

Civil war broke out on 11 August 1975 when UDT seized key installations in Dili and Baucau. Having obtained large quantities of Portuguese Army weapons, Fretilin won the bloody war and declared independence on 3 November that year. A week later Indonesia sent in a massive invasion force "to prevent the consolidation of a Marxist state on our doorstep". Timor-Dili was annexed and became Indonesia's 27th province.

Mountainous East Timor, with jungle-clad peaks rising to 10,000 feet, is ideal guerrilla country. In 1942-3 300 Australian commandos outfought 16,000 Japanese troops and killed 1,500 of them for a loss of 40 of their own men. Now Fretilin fighters took to the mountains to wage war against the occupying Indonesian troops. The conflict has gone on ever since.

The War in 1985

By 1985 more people had been massacred in the East Timor war than in any other contemporary war except Kampuchea. As early as 1979 some sources—including Australians who had once owned plantations in East Timor—claimed that

100,000 people had been massacred. This figure may then have been exaggerated, but in 1985 it was at least 100,000 and possibly 200,000.

Despite closing off access to human rights groups and to journalists, Indonesia has not been able to prevent exposure of its brutal tactics. Amnesty International, the Catholic Church and Tapol (the British Campaign for the Defence of Political Prisoners and Human Rights in Indonesia) have published well-documented evidence of widespread torture, murder, rape and imprisonment in East Timor.

At Kraras, when guerrillas killed 15 Indonesian soldiers, the army murdered 200 people, including women, children and the sick by setting fire to their houses. When the rest of the town's population fled to the mountains, the Indonesians mounted "Operation Clean Sweep". By combining encirclement with aerial bombing, the soldiers killed another 500 people.

In August 1985 a spokesman for Fretilin told the UN Special Committee on Decolonisation that Indonesia had succeeded in "subjugating" East Timor using French helicopters and tanks, British aircraft, American transport helicopters and counter-insurgency aircraft and West German and Dutch submarines and frigates. British governments routinely abstain when the self-determination of the East Timorese is discussed at the UN; Indonesia is a large market for British goods.

General L. B. Moerdani, Indonesian Army commander in East Timor, estimates (according to a UN working paper) that only a few hundred members of Fretilin remain in the mountains. This is in sharp contrast to Fretilin spokesman Abilio Araujo, who puts Fretilin's strength at 6,800, with units present in 80% of the country. US intelligence sources say Fretilin has only 600 men, but that they are among the best guerrilla fighters in the world.

Abilio Araujo has accused Indonesia of using chemical weapons against the guerrillas and the civil population, including defoliant spray and toxic gas. One of the most reliable witnesses of Indonesian ruthlessness is the senior Roman Catholic official in East Timor, Monseigneur Carlos Ximenes Belo. In a document smuggled out in 1985 and published in several Western newspapers, Mgr Belo stated: "An upheaval of gigantic and tragic proportions is taking place in which the identity of the Timorese people is threatened by destruction. Successive, systematic and regular clean-up operations are launched by the Indonesian army against centres of resistance. A war that continues for 10 years cannot be imputed to the blind obstinacy of a minority. The attempts to Indonesianise the Timorese people represents a slow assassination of their culture. To kill their culture is to kill the people themselves".

Despite the permanent threat of summary execution of men and youths considered to be helpful to Fretilin, resistance continues. Indonesian troops are afraid to go on patrol in the mountains and there have been many cases of desertion. Reliable sources say that in the year to July 1985 more than 900 troops were killed or wounded and that more than 200 "went missing".

It is the latter category of casualty which alarms the Indonesian troops. The Fretilin guerrillas are experts at swooping on isolated posts and making off with the soldiers and all their equipment. Sometimes a single soldier will "vanish" from a file of troops on a jungle trail. Soldiers on local leave and enjoying a swim from a beach will similarly "vanish". The troops are convinced that they are drowned by guerrilla swimmers. An Indonesian colonel who visited a village to choose a girl as his villa "companion" was found castrated next day.

About 80% of Fretilin's arms, ammunition and supplies are stolen from the army; the other 20% is sent in by boat to rendezvous with guerrillas at lonely beaches. This material is believed to come from Vietnam.

Otherwise there appears to be no foreign military involvement in East Timor. On at least one occasion Fretilin sent agents to Australia to find former soldiers willing to be mercenaries, but the Timorese could not offer enough money. American interest was aroused by independent reports which suggested that Fretilin is not as Marxist as the Indonesians claimed and in reality is democratic. A small group of Americans, believed to be CIA agents, managed to get into the island and talk to guerrilla leaders, but no agreement on aid was reached.

The Indonesian Navy and Air Force have virtually sealed off East Timor to prevent departures and arrivals. Resident foreigners who leave are not allowed to return; others have been deported. According to priests, the Indonesians are committed to genocide, but, they say, it is still possible that Fretilin guerrillas will persuade the Indonesian government that the cost of trying to wipe out the Timorese is too high.

El Salvador Civil War

El Salvador Civil War

VICTORY BY EXHAUSTION

Background

This war has been so confusingly presented in the Press that it can make no sense without a reasonably elaborate background briefing. Many groups and personalities are involved, but it is possible clearly to present them, their policies and the military operations of the war.

The conflict, which began in 1980, is generally considered to be a civil war, but it is more clearly a guerrilla war with five groups of rural guerrillas fighting the army. El Salvador, with an area of 210,000 km^2, is an ideal country for guerrilla warfare; much of it is mountainous and jungle-covered, with thousands of rural villages where guerrillas can seek shelter. The army cannot effectively use its numerical superiority and rarely finds a chance to employ its armour.

The guerrillas oppose the right-wing government of José Napoléon Duarte, whom many foreign correspondents regard as a fascist.

There is no doubt about the ideology of the Salvadoran guerrillas who oppose him; they are confessedly Marxist. The guerrilla movement was not united until 1980 when its leaders met several times in Havana, Cuba, at the invitation of Fidel Castro. In exchange for Cuban promises of more military aid, the guerrillas formed three organisations. They were:

(i) The Democratic Revolutionary Front (FDR), an amalgam of revolutionaries and representatives of alienated left-wing parties whose job is to plead the guerrilla case abroad.
(ii) The Unified Revolutionary Directorate, a fifteen-member war council of top guerrilla leaders–*commandantes* as the Salvadorans call them.
(iii) The Farabundo Marti National Liberation Front (FMLN), a co-ordinating and organisational body for the guerrilla groups.

The pact evolved in Cuba was supposed to end years of bickering which prevented the guerrillas from being nothing more than a nuisance to the Duarte régime. Instead, it created a new crisis around the two most powerful rebel organisations—the 3,000-member Popular Liberation Forces (FPL) led by Salvador Cayetano Carpio and the 4,000-member People's Revolutionary Army (ERP) of Joaquin Villalobos. The dispute was resolved in a typically Salvadoran way. In 1983 Carpio died under mysterious circumstances and his group splintered. Villalobos emerged as "first among equals" in the guerrilla hierarchy.

Carpio's influence as a guerrilla leader had been profound and he was often called the Ho Chi Minh of the Salvadoran revolutionary movement. By 1947 he had joined the illegal Salvadoran Communist Party and eventually he became its general secretary. Fanatically secretive, he was the chief exponent of "prolonged popular warfare", a hard-line strategy that would in his view probably culminate in direct confrontation with the US. Increasingly, it conflicted with the view officially presented by the other FMLN members, that the revolutionaries would negotiate a vague power-sharing agreement with the Salvadoran government.

Villalobos, only 37 when the 63-year-old Carpio died, has long been the leading guerrilla exponent of such a strategy. As a teenage student leader he began in 1971 to form clandestine left-wing groups to foster his own version of "armed struggle". His units later specialised in kidnapping and urban terror. Villalobos created a major scandal in 1975 when he ordered the execution of a well-known Communist poet, Roque Dalton Garcia, on trumped-up charges of being an agent of both the CIA and Cuba. In reality, Garcia was Villalobos's chief political rival. The killing led to bitter internal fights and schisms.

Villalobos's brand of revolutionary pragmatism appealed to younger Salvadoran rebels, who flocked to his banner. It was also accepted by Carpio's deputy, Melida Anaya Montes, a formidable female fighter, and Carpio's junior commanders.

Less than 3 months after a meeting to weld together Carpio's old party and the Villalobos group, Anaya Montes was stabbed to death in the Nicaraguan capital, Managua. Just who ordered her killing is uncertain, but many people believe it was Villalobos, who feared that the older Mrs. Montes, aged 52, would want to dominate him.

There are three other FMLN chieftains. One is Shafik Jorge Handal, general secretary of the Salvadoran Communist Party and commander of its armed wing, the 1,500-member Armed Forces of Liberation. The son of Palestinian immigrants, Handal is the organisation's main go-between with the Soviet Union and Cuba. Because of his standing in Moscow, he has great influence in guerrilla councils.

Next in line is Eduardo Sancho Castaneda, leader of the 2,000-member Armed Forces of National Resistance (FARN). He is better known by his *nom de guerre*, Ferman Cienfeugos. A founding member of Villalobos's group, Castaneda broke away after the Garcia murder. Ideologically, FARN is believed to be the most nationalistic and the most hostile to Soviet and Cuban influence. Least influential is Roberto Roca, who leads the 300 members of the Central American Workers' Revolutionary Party.

FMLN units have a well-organised anti-government propaganda programme. The guerrillas run two newspapers and three radio stations that broadcast out of Managua and Mexico City. Salpress, a guerrilla news service started in 1980, has correspondents in nine countries and distributes weekly summaries and features to newspapers worldwide. As a result, in the US 347 "solidarity committees" collect money for the movement. In 1983 the guerrillas hired a US public relations firm with offices in Manhattan and Washington to present their case.

President Duarte, a Christian Democrat, resorted to death squads to curb guerrilla activity and to deter ordinary people from giving assistance to the FMLN. The death squads consist mainly of off-duty members of the two main security forces—the 2,500-strong Treasury Police (whose nominal job is to collect taxes) and the National Guard, which also has 2,500 men.

Between 1978 and 1984 the death squads murdered 14,629 peasants, 2,255 industrial workers, 1,783 students and 25,789 others. These are the statistics compiled by probably the only group capable of doing the job—Socorro Juridico, the legal aid department of the Roman Catholic archbishopric of Salvador. It is likely that some of these killings were carried out by the army in anti-guerrilla sweeps.

During the period 1982-4 artillery and air bombardment was widely used to soften up what were supposed to be guerrilla targets. In fact, these attacks were mostly against civilian settlements and they caused widespread casualties.

In October 1983 Duarte submitted himself to the first round of a democratically organised presidential election. His main opponent was former army major Roberto d'Aubuisson, himself regarded as a death squad "godfather". Duarte campaigned on a policy of "doing away with the death squads", of strengthening his "kindergarten democracy" and of amnesty for those guerrillas who wanted to join Salvador as "free citizens in a liberal society".

An element of desperation and ferocity entered the armed forces' tactics in May 1984, when the Salvadoran Air Force used incendiary bombs against the civilian population in the zone around the Guazapa Volcano; one type of bomb resembled napalm and the other white phosphorus.

It seemed that the Salvadoran army of 46,000 was steadily overcoming FMLN resistance. At this point, 23 October 1984, one of the army's best combat officers, Lieutenant-Colonel Domingo Monterrosa was killed when an American-supplied Huey helicopter crashed at Joateca. The guerrillas' clandestine broadcasting station, Radio Venceremos, claimed that it had been shot down by machine-gunners. Whatever the cause, the effect was devastating. Much of the basis for the claim that the Salvadoran Army was doing better was provided by Colonel Monterrosa. Largely by personal example, he produced a new aggressiveness among the troops he commanded.

Monterrosa also had a contempt for danger. He was known to lead a squad in a charge against a guerrilla position; sometimes he drove back to his headquarters in the city of San Miguel in the risky after-dark hours. His fearlessness may have been his undoing. The town of Joateca had been turned into a forward command post for an operation Monterrosa had launched to clear the guerrillas out of their strongholds in northern Morazan province. The colonel had been flying in and out for 4 days before his crash.

Whether the helicopter was shot down or just fell, the colonel's death was a reminder of the vulnerability of the helicopter in modern warfare.

The War in 1985

The year began on a violent note. On 6 January, in the Union region—in the extreme east—Pedro Rene Yanes, one of Duarte's closest advisers, was murdered. In a single week several other officials were killed, supposedly by Major d'Aubuisson's death squads. Jose Simon Parada, a leader of Popular Democratic Unity, was murdered by three unidentified killers who waited for him outside his home on the northern outskirsts of San Salvador.

A bomb was planted in the home of the mayor of Quezaltepeque—like Duarte a Christian Democrat—and the mayor's 18-year-old daughter was killed. Murders had dropped in 1984 compared with 1983 (from 1,259 to 200), but the priests said that this was because there were fewer people left to kill.

On 20 January guerrillas blew up the bridge across the River Torola, thereby isolating the Morazan region from the rest of the country. The army's penetration of the rebel-dominated sections of the countryside led the guerrillas to begin a new campaign of urban terrorism, much like the one they pursued in the early days of the war. Rebel leaders warned that an attack on 19 June on streetside cafes in San Salvador was the beginning of a "war of attrition" against the government and that it would "bleed enemy forces". Thirteen people, including six Americans, died in the terrorist attack.

Such tactics can backfire. The cafe slaughter was condemned by many Salvadorans from all parts of society. It also produced strains between guerrilla leaders and their allies in the Democratic Revolutionary Front, the umbrella organisation for leftist political parties.

The guerrillas attempted a few large-scale operations. For instance, about 400 attacked a protected government communications outpost on the slopes of the San Salvador volcano. The tiny hamlet of Santa Cruz Loma, 33 miles from San Salvador, was attacked by 200 rebels. They killed 17 villagers, including 6 members of the local civil defence section, as well as a woman and 3 small children. The village civil defence men had automatic rifles but no radios to call for help from an army post only 5 miles away. Under the government's new civil defence concept, the government is relying on 7,000 better prepared volunteers to provide a visible, continuous security presence in about 60 towns and villages.

The guerrillas had a setback when rebel leader Americo Mauro Araujo, second in command of the Salvadoran Communist Party, was captured. Under torture, he gave the government much information about the guerrillas' clandestine operations.

During 1985 the rebel forces were beginning to feel the effect of 18 months of aggression by the armed forces. With American help—$454 million in 1985—the army increased its strength to 50,000. In addition, the air force was strengthened by 9 A-37 jet bombers, 2 AC-47 gunships and 30 rocket-firing helicopters. This superior strength left the rebel organisations of FMLN with little choice but to step up even further their urban terrorism.

Villalobos, by small hit-and-run attacks, believes that he can exhaust government and troop morale. He told visiting correspondents that his strategic aim was to overcome the capacity of the Reagan administration to continue supplying the Salvadoran army.

Less dramatic but no less effective is the guerrilla's long-standing campaign of economic sabotage. For instance, in September and again in October 1985 rebels set up road blocks that stopped traffic on most of the country's main roads and destroyed electricity power lines—by blowing up 1,000 pylons and poles—in several provinces. The government can ill afford the cost of repairs to damage caused by the guerrillas. In September Duarte decreed a "war economy", a state of austerity with tough controls on profits and prices.

The army's infantry faced ever greater dangers as 1985 progressed, with the guerrillas stepping up the use of landmines and booby traps; in one month the army had 682 casualties, 146 of them in landmine explosions. A Salvadoran army officer said: "The enemy has learnt that to put five soldiers out of action by blowing off a foot, a leg or a hand is much better than killing one man. The cost to us is tremendous".

The army has problems too in winning popular sympathy. Following complaints in 1984 about indiscriminate bombing, Duarte issued regulations requiring that the military high command must authorise all air attacks on rebel targets. But while the restrictions reduced excesses, many reports of military abuses continued. On 6 August air force planes wrecked the hamlet of Ocotal in the province of Chalatenango, killing some women and children.

There were real fears late in 1985 that the Salvadoran war might spill across the country's eastern border into Honduras. Honduran soldiers entered a UN camp for Salvadoran refugees in Honduras, killed some people, wounded others and abducted ten. Honduras military officials claim that the UN camp is a haven for Salvadoran rebels. International relief workers say that it is populated only by refugees. Following this incident, the FMLN *commandante* Shafik Handal threatened to "regionalise the conflict" by attacks into Honduras.

The American administration has throughout the long war supported and aided the government, whatever its shortcomings, because the US considers that a Marxist government in the region would bring chaos to all of Central America. With American advisers and instructors, the Salvadoran Army is well organised and equipped. It has 4 infantry brigades (9 battalions in all), 14 independent infantry battalions, 1 paratroop battalion and 3 special forces battalions—the Atlacatl. In addition, there is 1 artillery brigade, 1 engineer battalion, 1 anti-aircraft battalion and 1 armoured brigade. The armour includes 12 AMX-13 light tanks, 18 AML-90 armoured cars and 30 armoured personnel carriers. The air force has 59 combat aircraft and, more importantly for counter-insurgency operations, at least 50 Huey UH-1H helicopter gunships.

The American instructors have induced the army command to switch to small patrols which go into the countryside for long periods and keep the guerrillas on the move. The brunt of the fighting is borne by the Atlacatl brigade and the Ramon Belloso Brigade; the "old army" is being deployed to guard strategic installations.

American-supplied C-47s equipped with three machine-guns made their first appearance in 1985. More of these aircraft, which cost $1.5 million each, have been ordered. The army's reliance on air power to soften up the guerrillas arouses comparisons with Vietnam.

The government is trying new propaganda tactics aimed directly at the rebels. Air force observation planes frequently drop thousands of conciliatory leaflets over guerrilla territory. One shows a photograph of two teenage rebels, named "Armando" and "Chando", who came down from the hills. They are smiling and drinking—according to the text—ice-cold beer. But it will take more than beer to end Salvador's battles.

The offer of safe conduct for guerrillas who give themselves up has met with a poor response. The army now issues a pass to guarantee "the physical and moral integrity" of defecting guerrillas, signed by the chief-of-staff, Colonel Blandon. The army is also offering rewards to guerrillas who hand over arms; they range from $600 for a mortar to 5 cents for a clip of ammunition.

At the end of 1985 Duarte and his armed forces appeared to be slowly winning the war, but the FMLN is the most potent insurgent army in Latin America and its field commanders have been attending training courses in Vietnam, Bulgaria and the Soviet Union. The FMLN's capacity for disruption and sabotage is endless.

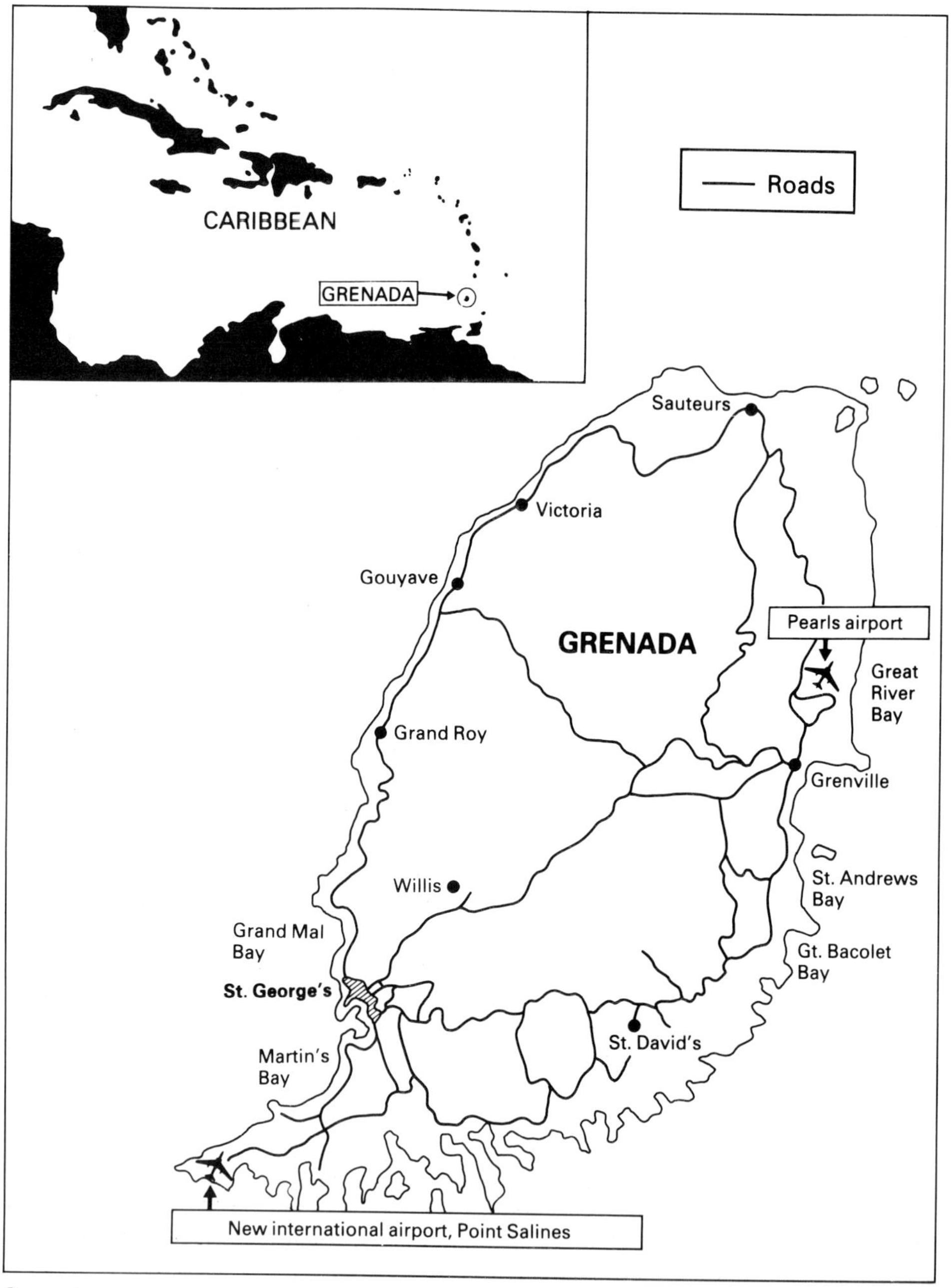
CARIBBEAN
GRENADA
Roads
Sauteurs
Victoria
Gouyave
GRENADA
Pearls airport
Great River Bay
Grand Roy
Grenville
Willis
St. Andrews Bay
Grand Mal Bay
Gt. Bacolet Bay
St. George's
Martin's Bay
St. David's
New international airport, Point Salines

Grenada

Grenada

MASSIVE "RESCUE MISSION"

While this war of October 1983 falls outside the period of conflict covered by Brassey's *War Annual I*, it is included in the book because it reveals a good deal about strategic objectives in the Caribbean Sea and Central America. Because the American administration and the invasion high command made it virtually impossible for journalists to report the war as it happened, many major factual inaccuracies appeared in the media; this report sets the record straight.

Without an account of the American invasion of Grenada, it is not possible fully to describe American military involvement in Nicaragua and El Salvador. Similarly, the Grenada episode—a superpower invading an insignificant island—is not comprehensible without reference to American military involvement in the Lebanon. On 23 October, just two days before the American invasion of Grenada, Muslim "Holy War" suicide attackers had killed 241 Marines with a single car bomb in Beirut. The US administration felt militarily impotent and politically humiliated. President Reagan needed a military triumph, and Grenada provided it.

The President, in an address to the nation, gave three reasons for the US invasion:

> "First, and of overriding importance, to protect innocent lives, including up to 1,000 Americans, whose personal safety is, of course, *my* paramount concern. Second, to forestall further chaos. And third, to assist in the restoration of conditions of law and order and of governmental institutions to the island of Grenada where a brutal group of leftist thugs violently seized power, killing the Prime Minister, three cabinet members, two labour leaders and other civilians, including children."

In 1979 left-wing revolutionaries of the New Jewel Movement, led by Maurice Bishop, overthrew the corrupt administration of Sir Eric Gairy. Having seized Fort George (originally built by the French in 1705 and named Fort Royal by the British), Bishop renamed it Fort Rupert. Rival revolutionaries led by General of the Armed Forces Hudson Austin later placed Bishop under house arrest, but on 19 October 1983 thousands of Grenadians demonstrated in his favour. General Austin ordered his troops to fire into the crowd, killing about 100 of them. Within an hour, Hudson had Bishop and his closest supporters executed in the fort.

Grenada had established a close political relationship with Cuba and the Soviet Union. This alarmed the Reagan administration, which saw its hegemony over the Caribbean threatened. The Russian leaders were pleased with the opportunity to embarrass the Americans, as they had done with their actions in Poland and Afghanistan, q.v.

The American strategic objective was to end the left-wing dominance in Grenada and to demonstrate to all governments within the Caribbean and Central America the military might possessed by the US. President Reagan ordered an invasion of Grenada on 25 October on the basis of contingency plans drawn up months before and waiting for just such a opportunity. The Americans would be invading a country which was part of the British Commonwealth and with a governor-general appointed by the British Crown.

The Task Force, under Rear Admiral Joseph Metcalf, consisted of 13 ships and 15,000 troops, including Marines, Rangers and Infantry. The principal ships were the USS *Guam*, 18,300 tons, the flagship, which carried 20 helicopters; the USS *Saipan*, 39,000 tons, with 26 assault helicopters; and the USS *Independence*, 79,000 tons, with a crew of 4,940 men and carrying aircraft. The main land-combat formation with the fleet was the 22nd Marine Amphibious Unit (Lieutenant-Colonel R. Smith) of 400 marines equipped with helicopters and assault craft capable of landing 5 M-60 tanks, 13 amphibious armoured vehicles (LVTP-7) and jeeps with anti-aircraft missiles. In Barbados, about an hour's flying time distant, C-130 Hercules were waiting to support the invasion by dropping 500 Rangers.

The massive US Task Force was augmented by groups of soldiers and police from Antigua and Barbuda, St. Lucia, Dominica and St. Vincent, Jamaica and Barbados. Numbering only 300, they had no military significance; politically they were important, because they gave apparent legitimacy to the American invasion.

Opposing the Americans was the Grenadian People's Revolutionary Army, the Grenadian militia and some Cubans. The PRA consisted of about 1,000 troops led by General Austin, a former prison warder, and trained by 40 Cuban instructors permanently on Grenada. Equipment had been provided mainly by the Soviet Union and European states and consisted of AK-47 and AKMS automatic rifles, 80-mm anti-aircraft weapons, rocket launchers and 7 eight-wheel BTR60 Soviet armoured personnel carriers. The Grenadian militia numbered no more than 2,000, and its members had received only rudimentary training, because General Austin feared that it could become more powerful than the army itself.

The Cubans numbered 784, and most were engaged in building the Point Salines airport, a major work of construction which the Americans claimed could be used as a Cuban–Soviet base for launching air attacks against US targets. Only 45 of the Cubans were military personnel. The others had been taught, as most Cubans are, how to use a rifle, but they were in no sense professional soldiers. All 784 came under the command of Colonel Pedro Tortolo Comas, Chief of Staff of Cuba's "Army of the Centre", who arrived the day before the American invasion.

The American strategy was to overwhelm the defenders by sheer weight of numbers. The tactical plan was to land Marines by helicopters and to airdrop the Rangers simultaneously. The invasion began between 6 am and 7 am on 25 October. Men of the 22nd MAU were helicoptered to Pearls airport on the western side of the island; one Cobra helicopter was shot down. At the same time the 500 Rangers dropped from 10 Hercules onto Point Salines airport, where Colonel Tortolo led a spirited resistance by PRA troops, militiamen and Cubans. With Point Salines reasonably secure, troops of the 82nd Airborne Division were flown in from Fort Bragg, North Carolina.

The Americans' main objective at Point Salines was the capture of the True Blue campus at St. George's University Medical School, in which about 30 American

Men of the US 82nd Airborne Division at the Grenada Beach Hotel, Grand Anse, October 1983.

This photograph of youths in the Grenadian Army was taken just before the war. Some boys in this squad were only 9 years of age. Even with Soviet AK-47 rifles they had little more than nuisance value as soldiers.

students were staying. The US command had feared that the Cubans or Grenadians would seize them as hostages.

As the assault troops went in, broadcasts from Barbados urged members of the PRA to surrender; propaganda leaflets and safe conduct passes were dropped from the air and thrown from American vehicles. This material had been prepared by the US Army's Psychological Operations Battalion,* a platoon of which was quickly ashore with the Marines.

Throughout the first day, American helicopters and other aircraft raided enemy positions around St. Georges, Point Salines and Fort Rupert. The Grenadians responded with anti-aircraft and automatic fire and brought down another helicopter. During the afternoon American aircraft bombed and badly damaged Crazy House, the Grenadian insane asylum, and killed about 40 inmates.

A truce seemed possible that night, and in Havana it was widely believed that fighting had ceased. Local people thought that both sides had stopped fighting, but next day, the 26th, American troops closed in on St. Georges. Marines were helicoptered to Grand Mal Bay, just outside the city, and three M-60 tanks blocked the road to the north. By now another objective had been achieved—"securing the person of the Governor-General Sir Paul Scoon". He was flown to the flagship for talks with Admiral Metcalf.

The last major engagement occurred late that afternoon, when Marines and Rangers seized the Grand Anse Campus of St. George's University Medical School, where hundreds more American students were sheltering. Helicopter gunships and Hercules attacked the enemy positions and the 82nd Airborne troops occupied the buildings. Colonel Tortolo, with 10 Cubans, was among the main enemy group surrounded at Point Salines, but he and his Cubans slipped away and later found refuge in the Soviet Embassy.

By Thursday, 5,000 paratroopers, 500 Rangers and 500 Marines—with the 300 Caribbean soldiers and police—were in control of Grenada. The populace at large welcomed the Americans' presence; local journalists rejected the label of "invasion" and announced that the US action was a "rescue mission" to save Grenadians from the oppressive rule of General Austin.

"We got there just in time", President Reagan said. He meant that the invasion had prevented the Cubans and Russians from using Grenada as an anti-American base, but nothing was found to indicate that such a base was being prepared. The war was popular in the US but the British government was horrified and angry; even the Prime Minister had not been consulted about the projected invasion of a British island reigned over by Queen Elizabeth II.

The American force had 18 men killed and 45 wounded. The Cubans lost 24 killed and 40 wounded, the Grenadians 60 killed and 184 wounded and several hundred taken prisoner. For several weeks some revolutionaries eluded a massive manhunt, but neither the regulars of the PRA nor the militiamen turned to guerrilla warfare, which would have been possible in the island's mountains and jungles. Nevertheless, the PRA had fought well—and better than the American command had anticipated. Most of the American troops left on 12 and 13 December and Major-General Jack Farris handed over military command to Lieutenant-Colonel Delroy Ormsby of the Jamaica Defence Force.

*It is not a normal battalion in strength and has only 200 members. Based at Fort Bragg, its task is to study and evolve techniques of propaganda for use in Latin America, the Caribbean, Africa and Asia.

In military terms, the conflict was inconsequential and revealed nothing about American combat leadership or the quality of US troops. However, it did show the army's hunger for medals. It issued more than 9,000 decorations—as distinct from a campaign medal—2,000 more than it had soldiers engaged in the war. The Navy and Marine Corps issued only 17 medals, all of them Purple Hearts, the standard award for a wound.

Hugh O'Shaugnessy, a British journalist who saw as much of thc conflict as anybody and who wrote a book about the war, claims that three conclusions can fairly be drawn about the invasion. "It was illegal; it was unnecessary as far as the rule of law in Grenada was concerned and it set back the cause of political democracy and long-term economic development in the region."

Guerrilla War in Guatemala

STATE TERRORISM

Background

Guatemala was once a tourist paradise—"the land of eternal springtime"—and it is still the richest and most industrialised country in Central America. It is also a deeply divided country: wealth and power are in the hands of the Ladinos—non-Indians of Spanish descent—who comprise roughly 40% of the population. Many Ladinos are shopkeepers or landowners. The majority of the population—descendants of the Maya Indians—have little political power, very little economic power and no social influence or standing.

Successive military dictatorships have suppressed the simmering discontent. The country has had military governments since a coup in 1954 that ended a decade of democratic government.

The military rulers used violence and terrorism to fight the leftist guerrilla movements. Even the moderate opposition, the Christian Democratic Party, has suffered from state terrorism; 560 of its workers were killed between 1981 and 1984. "Official" violence became so extreme that in 1977 President Carter stopped all US military aid in an effort to make the army leaders more reasonable.

The period 1982-3, during the regime of President Rios Montt, was probably the worst. The army had operated openly, bombing and killing supposed left-wing sympathisers, but that changed under President Montt. Throughout the country, the army conscripted about 350,000 men—from a population of 8.2 million—into paramilitary civilian defence patrols, with the ostensible purpose of protecting towns and villages from guerrilla attack. Patrol duty is obligatory in rural areas for men over 18, and those who refuse it are executed. Villagers serve a 24-hour shift on patrol every 8 days.

In some areas the patrols are greatly feared as a vigilante force in the service of the army. Two men who had left the town of Joybaj without the civilian patrol's permission were captured, tortured and decapitated. Their heads were presented to the army.

President Montt ordered a new-style anti-guerrilla "cleansing". An army unit enters a village with lists of suspected left-wing guerrilla sympathisers. The soldiers put a hood on one man from the village and make him walk among the people, whom he sees through eyeslits, and point out suspects. The village defence patrol is then ordered to execute them. The men are held down, their heads are yanked back by their hair and are then hacked off with machetes. In all such cases the heads are presented to the local military commander as proof that execution has been carried out.

During Montt's regime the army massacred 10,000 unarmed civilians. At a small village near Nahula in Totanicatan, soldiers discovered a cache of weapons and

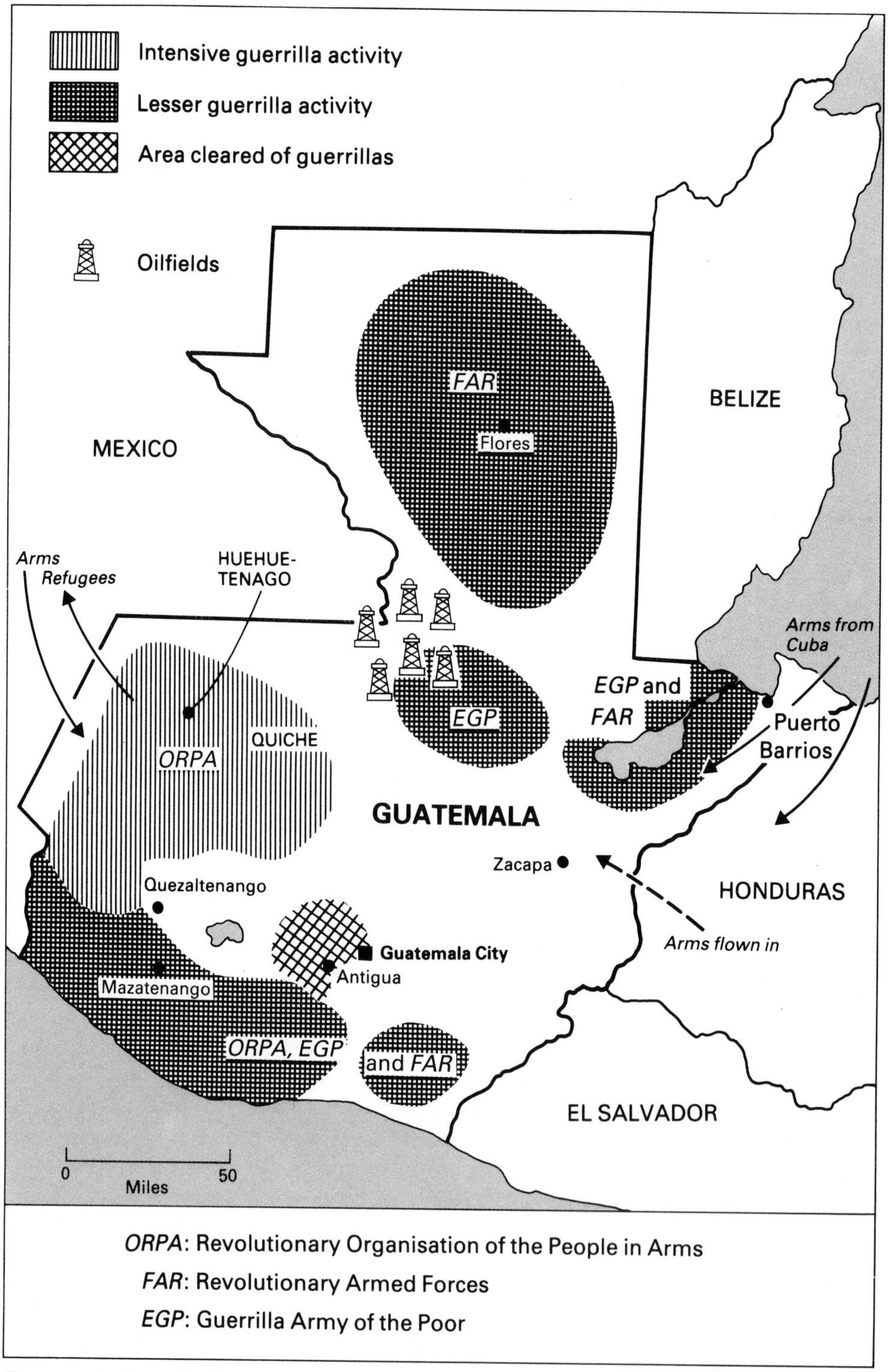

Guatemala Guerrilla War

slaughtered 300 villagers. In many cases the civilian defence patrols were used as the killers. A patrol from Chiul in El Quiche was taken to the village of Paraxtut and ordered to execute all the 335 adult males. The soldiers then raped the youngest women and killed all the women and children except one woman and two children, who escaped. Catholic nuns in Huehue-Tenago reported to the Vatican that the army was implementing a government policy of genocide.

During Montt's regime the army numbered only 18,000 men with inadequate equipment. Its artillery was largely confined to grenade-throwers and its small tanks were useless in most of the terrain where the fighting was taking place. The biggest arms suppliers in 1982-3 were the Israelis; Guatemalan opposition leaders alleged that the US was laundering military aid via Israel in order to avoid the need for congressional approval. The guerrillas, 9,000 strong, were well equipped, with Soviet-made RPG-7 grenade-launchers, Israeli-made Uzi machine-guns and M-16 rifles. Some of the supplies came from neighbouring Mexico. Other material arrived by boat from Cuba and was dumped on the shores of the Gulf of Honduras. Light equipment from Nicaragua came across Honduras by air and road.

The War in 1985

In 1985 the country was under the rule of yet another general, Oscar Humberto Mejia Victores, who succeeded in regaining American aid. The US administration of President Reagan was eager to retain an anti-Communist ally in Central America, and Congress authorised $45.5 million in aid. Regular military assistance was made contingent on the holding of democratic elections in November 1985 and on an acceptable human rights record.

The Victores regime has no such record. Violence is still widespread and institutionalised. For instance, two leading members of Guatemala's only human rights organisation, the Mutual Support Group, were murdered and two others fled the country. More than 100 Indians disappeared or were killed in the area around Patzun. Eight faculty members and scores of students disappeared from the National University. Several Indian social workers were kidnapped from the Roman Catholic University. The security forces consider all educators to be radicals. Six American linguists working on a US Aid programme were murdered in 1985.

Labour leaders, Roman Catholic priests, university professors and lecturers, and journalists are traditional army targets in Guatemala: 50 journalists have been murdered since 1977.

President Victores further developed the system of "model villages", similar to the strategic hamlets of the Vietnam War. Tens of thousands of villagers are forcibly resettled in areas of strong army control, to prevent them from aiding the guerrillas. Nevertheless, during 1985 the number of guerrilla attacks increased significantly in provinces on the Guatemalan–Mexican border. The guerrillas took on larger government forces, evidence of greater military capability.

Early in 1985 the four groups formed a coalition known as the Guatemalan National Revolutionary Unity. Co-ordination has improved considerably since then. Overall, the four organisations had fewer men in 1985 than before—perhaps only 3,000—but they were better armed and led. The strongest is the Organisation of the Armed People (OAP), with its headquarters on the southern slopes of the

Atitlan volcano, 2 days' march from Guatemala City across mountains and through tropical forest.

OAP's chief is Rodrigo Asturias, whose father, Miguel Angel Asturias, won the Nobel prize for literature in 1967. The guerrillas live in muddy, uncomfortable conditions, but believe that such a life keeps them fit and strong for their "national revolution". The OAP fighters are well armed, particularly with Israeli-made weapons captured from the Guatemalan Army.

Asturias has almost abandoned ambush as a tactic. He has come to believe that no ambush can cause enough casualties, inflict enough damage or lead to the capture of enough equipment. In place of ambushes, he is content to mine jungle tracks and to kill the occasional soldier by booby-trap.

His tactics have changed to attacks on army bases, outposts and supply depots. Each attack is thoroughly planned and every assault during 1985 was achieved with surprise. This is a tribute to OAP's intelligence officers who cleverly mislead the army into expecting a raid at one point, only to have another one hit.

Having become adept at night-fighting, the OAP men usually attack during the hours of darkness and, to achieve surprise, they have been known to make a raid during a tropical rainstorm. The objectives vary: sometimes the aim is merely to capture supplies. On several occasions either OAP or one of the other groups has raided a base for no other purpose than to capture the commander. The guerrillas believe that this is the greatest blow to morale which a unit can suffer.

Guatemala ranks as the most dangerous place in Central America in terms of being dragged out of a house at night or snatched off the streets during the day by "uniformed unidentified men". The government's explanation is always that murderers and kidnappers in army or police uniforms are guerrillas dressed in the official uniforms. Another line to account for Indian deaths is that Indians die as the result of drunken machete fights or old feuds. The average number of violent deaths per week in 1985 was 190, an increase of 40 a week since the time of President Montt.

How much popular support the guerrilla groups have is difficult to assess, but there seems to be great sympathy for their cause. But not even the most optimistic leader would claim that victory is close.

Diplomats and other political analysts do not believe that the elections will change the human rights situation. They say that the army, now 33,000-strong, will remain the real power in the countryside and that counter-insurgency demands will over-ride all other considerations.

Guatemala claims sovereignty over neighbouring Belize, a British-protected state. There is no doubt that the army, if it could end the guerrilla war, would want to invade Belize, where Britain maintains 1,800 troops as a deterrent.

In January 1986 a civilian, Vinicio Cerezo, became president. A Christian Democrat, Cerezo had already survived three assassination attempts by right-wing death squads. In Guatemala City people desperately wanted to believe that their new civilian president would be able to wrest power from the generals, establish independence from the United States and bring to an end the 31-year reign of terror. It had been responsible for 100,000 deaths and 88,000 "disappearances".

Holy War—Jihad

WORLD-WIDE STRUGGLE

Wars fought in the name of religion are as old as history itself, and Islamic holy war—or *jihad*—has existed as an institution since the time of the prophet Muhammad in the seventh century. The first *jihad* took place at Badr, near Medina. Muhammad fought seven *jihads* in Africa, the Persian Gulf and the Middle East.

For the modern Western world *jihad* became a reality rather than philosophical abstraction or a subject of academic study in 1979.

In that year Ayatollah Khomeini declared *jihad* against the West—the US in particular—and against other "enemies of God", including Egypt and its president, Anwar Sadat. When on 4 November 1979 Iranian Revolutionary Guards seized the American Embassy in Teheran and held the diplomats hostage (for 400 days), Khomani was embarking on a stage of *jihad*.

Translated, *jihad* means "a great striving" in the name of Allah to bring the world under Islamic rule.

Jihad is not necessarily "declared" in the Western understanding of the word, but it is genuinely a "total war". While it means armed struggle and battle, it equally means a great striving through economic and political pressures, through subversion and propaganda, through conversion of non-Muslims and through penetration of non-Muslim societies. While *jihad* might be violent in one place, at the same time in another place it could be low-key and limited to propaganda. In this book we are more concerned with its armed and military manifestations.

Jihad is not merely Islam's war against the Christian West and against the Jews, it is also aimed at all Muslim monarchs, as monarchies are anathema to strict Shi'a Muslims. Other enemies are "reactionary" Islamic regimes, such as that of Egypt, and those which Muslims declared to have betrayed Islam, such as Morocco. For Shi'a Muslims, this category includes most Sunni Muslims. President Gaddafi of Libya, who regards himself as the Mahdi—the "expected one" of Islam—regards his opponents as enemies to be destroyed by force in holy war.

In December 1979 the attempt by 500 armed insurgents to take over the Great Mosque at Mecca was an event of holy war. In the fighting which followed, the Saudi guards killed more than 300 insurgents, including their leader, Muhammad al-Quraishi, who also regarded himself as the Mahdi. All the attackers taken prisoner were beheaded. The attack was holy or jihadic in the eyes of the conspirators, because the Saudi royal family, the guardians of the greatest of Muslim mosques, is alleged to be religiously corrupt. Had the raiders taken control of the mosque, then holy war might well have spread faster.

The assassination of President Sadat on 6 October 1980 was another act of holy war. With grenades and Kalashnikovs some of Sadat's own soldiers killed him and other people at a military parade to commemorate "the Great Crossing"—Egypt's attack against Israel in 1973. Sadat's offence was that he had become an enemy of

God by making peace with Israel, a nation of "infidels". It must be realised that to Islamic Jihad the attackers were not murderers but soldiers of Allah.

On 18 April 1983 the US Embassy in West Beirut was wrecked by a car bomb; 63 people, including 17 Americans, were killed. This was a static car bomb of the type used by terrorists in Northern Ireland, and the atrocity was regarded as just one more act of violence in a city where violence has been a way of life since 1975.

Before October 1983 the West—even many Western military analysts—had taken holy war to be a peculiarly intra-Islamic phenomenon manifested by the occasional murder of an Arab leader by a terrorist group.

On 23 October this perception began to change when a suicide attacker rammed a truck-bomb with 2000 lb of explosives into the headquarters of the five-storey building which housed the US Marines Headquarters in Beirut; 241 marines were killed. Minutes later a second suicide truck-bomb crashed into the French paratroops' base close by and 58 paras died. On 4 November 1983 another suicide bomber attacked the Israel Border Police HQ in Tyre and 60 people, most of them Muslims, were crushed to death in the explosion.

In December 1983 a truck-bomb badly damaged the US Embassy in Kuwait and five remote-controlled bombs blew up Kuwaiti installations. These too were actions of holy war. The Kuwait explosions gave clear warning that the oil emirates of the Gulf were in danger of being crushed between two holy wars—that directed against the US and its Western allies and that taking place between Iran and Iraq.

Because of the intensity of religious feeling which it generates, Islamic holy war has a momentum of vengeance which makes the world its battleground. The French learned this on New Year's Eve 1983 when an express train travelling from Marseilles to Paris had a carriage torn apart by a bomb. Another bomb damaged Marseilles railway station. The Islamic Jihadic Organisation claimed responsibility.

Books written by some Muslims frankly reveal the will for conquest. One of the most significant is *Jihad—a Ground Plan*, written by 'Abd al-Qadir as-Sufi ad-Darqawi and published in Britain, in English.

> "We are at war, [says ad-Darqawi] and our battle has only just begun. Our first victory will be one tract of land somewhere in the world that is under the complete rule of Islam. . . . Islam is moving across the earth. . . . Nothing can stop it spreading in Europe and America".

While in exile in Paris, Khomeini wrote:

> "Holy war means the conquest of all non-Muslim territories. Such a war may well be declared after the formation of an Islamic government. . . . It will then be the duty of every able-bodied adult male to volunteer for this war of conquest, the final aim of which is to put Koranic law in power from one end of the earth to the other".

By prescription and tradition, holy war is war without scruple, and it is fought until victory is achieved. Truces may be arranged, but their purpose is to gain some advantage, such as time. A treaty with an infidel may be legitimately broken. In older times, even when a jihadic army was on the battlefield, face to face with an infidel army, the unbelievers could save themselves by embracing Islam. Islam's enemies may still do this. But if they fight, they cannot expect their lives to be spared should they be taken prisoner; there is no such provision under Islamic law.

The Prophet Muhammad divided all men and all communities into just two groups—Dar al-Islam (the abode or house of Islam) and Dar al-Harb (the abode of war). In Dar al-Harb are all those who reject Allah and his Messenger; they are to be considered in a state of enmity against those of the true faith, Islam. The distinction has never been amended or suspended.

The United Assembly of Islam is the political co-ordinating centre for the spread of Islamic fundamentalism, but its active operations are run by the Jihad Commission, whose activities first became known in 1982. A entire organisation exists to plan holy war and to co-ordinate activities, which in the West are mostly confined, for the present, to terrorist activity. The central organising body has three sub-units.

1. The Hezbollah, or "Army of God".
2. Pro-Khomeini Shi'as.
3. The Pasdaran or Special Revolutionary Guards.

These networks engage in two types of terrorism. First is "Red Terror", which includes assassination, sabotage and brainwashing-indoctrination. Second is "White Terror", which covers everything that the Red does except brainwashing. The White Terror network includes groups from the Shi'a Amal in Lebanon, the Syrian-backed PLO breakaway faction of Abu Mousa, several black Americans and the group organised by the Palestinian terrorist, Abu Nidal.

The Jihad Commission gives ideological direction to holy war. These are the principal members of the Commission, generally known only by code-name:

Chief of the Commission is a man code-named Mirshashem, about whom little is known.

Hashemi Rafsanjani, Speaker of the Iranian parliament.

A senior member of the Iranian Joint Chiefs of Staff.

Hojatosalam Mohsen Husseini Khoeinia. He is Khomeini's personal representative to Mecca. Trained in Leipzig, he is a member of the "Islamic Action Group", founded in 1984.

Moussavi Khamenei, President of the Islamic Republic of Iran.

Ali Akhbar Velayati, Iran's foreign minister.

Hoj. Soyyed Muhammad Khatemi, Iran's propaganda minister.

A Pasadaran leader who is a former agent of Savak, the Shah's intelligence service.

An ayatollah who represents the Savama, Khomeini's intelligence service.

Ayatollah Janati, chief of the organisation called Council of the Fighting Clergy. This group directs the thought and practice of all ayatollahs.

Soyyed Muhammad Bagher Hakim, former head of the supreme council of the Islamic Revolution of Iraq.

Abbas Zamani aka Abu Sharif, possibly a KGB man on the Commission. A founder of Hezbollah in Lebanon, he was head of Pasdaran and spent some time in Pakistan as Khomeini's adviser to President Zia al-Huq.

Decision-making for *jihad* in Lebanon is the responsibility of Sheikh Muhammad Fadlallah and his adviser, Seyed Jaffar Morteza, known as "the unseen man" in Lebanon. They receive their orders from Ayatollah Montazeri in Teheran.

In each country the work of the Commission differs. In Britain it is directed by the Islamic Foundation, whose couriers are believed to be Maltese. In France the

Iranian embassy directs terrorism and hires professional assassins and terrorists. For the murder in Paris of the Iranian opposition leader General Oveissi the Carlos and Abu Nidal groups were brought in for a co-operative effort.

While the strategy of *jihad* is direct, its tactics vary from country to country, enemy to enemy. This has led some Western commentators to the false conclusion that jihadic tactics are erratic, random and "mindless". In fact, every action is thoroughly planned for the maximum psychological, propaganda or military impact *at the particular moment.* Pistol close-quarter assassination of a public figure may be considered preferable to a time-bomb in his car; a suicide attack could achieve religious and political results which a remotely controlled bomb might not.

During 1984-5 these wars were partly jihadic:

The Iran–Iraq War (the so-called Gulf War). It is a holy war from Iran's point of view, though not from that of Iraq, though occasionally Iraqi spokesmen claim, for the sake of propaganda within the Islamic world, to be fighting a *jihad.*

The Bangladesh–Shanti Bahini War. The Bangladeshi Army's 24th Division claims to be fighting the guerrillas in the name of Islam.

Afghanistan War. The Afghan guerrillas fighting the Soviet troops and their Afghan Army allies have frequently asked the Islamic world for help "in the name of *jihad*". Saudi Arabia took this appeal so seriously that in 1984 it backed an attempt to set up a government in exile in opposition to the pro-Soviet Karmal regime. Because of differences between the moderates and the fundamentalists, as well as differences in the fundamentalist camp itself, the attempt failed.

Indonesia—East Timor War. Indonesian mullahs have said that the war being waged by government troops against the revolutionary forces in East Timor is *jihad*; this endorsement reduced the criticism against the war in Indonesia, which has the largest Muslim population in the world.

Lebanon Internal Wars. These conflicts are heavily jihadic in the intensity of conflict. The Lebanese Shi'as, together with their Iranian allies, consider they are fighting a holy war against the Christian Lebanese.

Philippine Internal Wars. Muslim guerrillas have been fighting an unremitting war against the government for 12 years, and ayatollahs throughout the Islamic world have declared their activities to be a *jihad.*

In the non-Muslim world holy war activities are most conspicuously directed against American targets. An Iranian source in Baalbek, Lebanon, claims that in 1984-5 the success rate for *jihad* actions against US targets was 92.5%. One major success was another suicide attack on the US Embassy in East Beirut on 27 September 1984 when 12 people were killed and 35 wounded.

Holy war is also directed against the French because of the French government's support for Iraq in the Gulf War and to a lesser extent against Britain and West Germany. In religious-political terms the chief enemy is Israel. (*See* Israel's Withdrawal from Lebanon.)

Despite many security precautions instituted during 1984-5, embassies and other public buildings are still vulnerable to attack by terrorists willing to sacrifice their lives. Effective counter-tactics have yet to be involved. The Western Intelligence agencies have found it virtually impossible to penetrate the *jihad* organisations. Retaliation is useless and probably counter-productive unless it is aimed precisely at the perpetrators of a terrorist act. This would indicate excellent intelligence, but if intelligence were so good, then the act might well have been prevented.

Some American experts—William Quandt of the Brookings Institute was one of them—in 1984 advocated closer ties with Syria as a possible way of reducing the risks of jihadic terrorism. He argues that neither Syria nor any other of the main powers in the area benefit from terrorist acts in the name of holy war and might be able to prevent them. This suggestion betrays an ignorance of the fervour which motivates *jihad.* Holy war planners cannot be dissuaded from their intentions by argument or even threat from their backers.

Extremes of Violence: New Tactics

In 1985–86 there were many indications that Holy War leaders were prepared to go to extreme lengths in their actions against the West. The seizure of the Italian cruise liner *Achille Lauro* was one such episode; the hijacking of a TWA aircraft to Beirut and the brutal killing of one of its American passengers was another. Yet a third was the hijacking of an Egyptair plane between Athens and Malta; the terrorists on this occasion murdered some Israeli and American passengers before the plane was stormed, at Luca Airport, Malta, by Egyptian commandos. The incompetent attempt at rescue left many of the passengers dead—as well as three of the four terrorists. The bloodiest holy war actions occurred at Rome and Vienna airports on 26 December 1985 when 15 people were killed and another 119 wounded by holy war attackers, Privately, Gaddafi claimed "credit" for authorising the massacres and publicly referred to them as "heroic operations by Palestinian martyrs".

It was learned during 1986 that Islamic *jihad* had decided to make kamikaze air attacks on American warships and certain US installations, such as embassies, in the Middle East. To this end other "martyrs", mostly Shi'a Muslim Iranians and Lebanese, were in training as pilots of light aircraft. Packed with explosives, the planes would be deliberately crashed onto the selected targets.

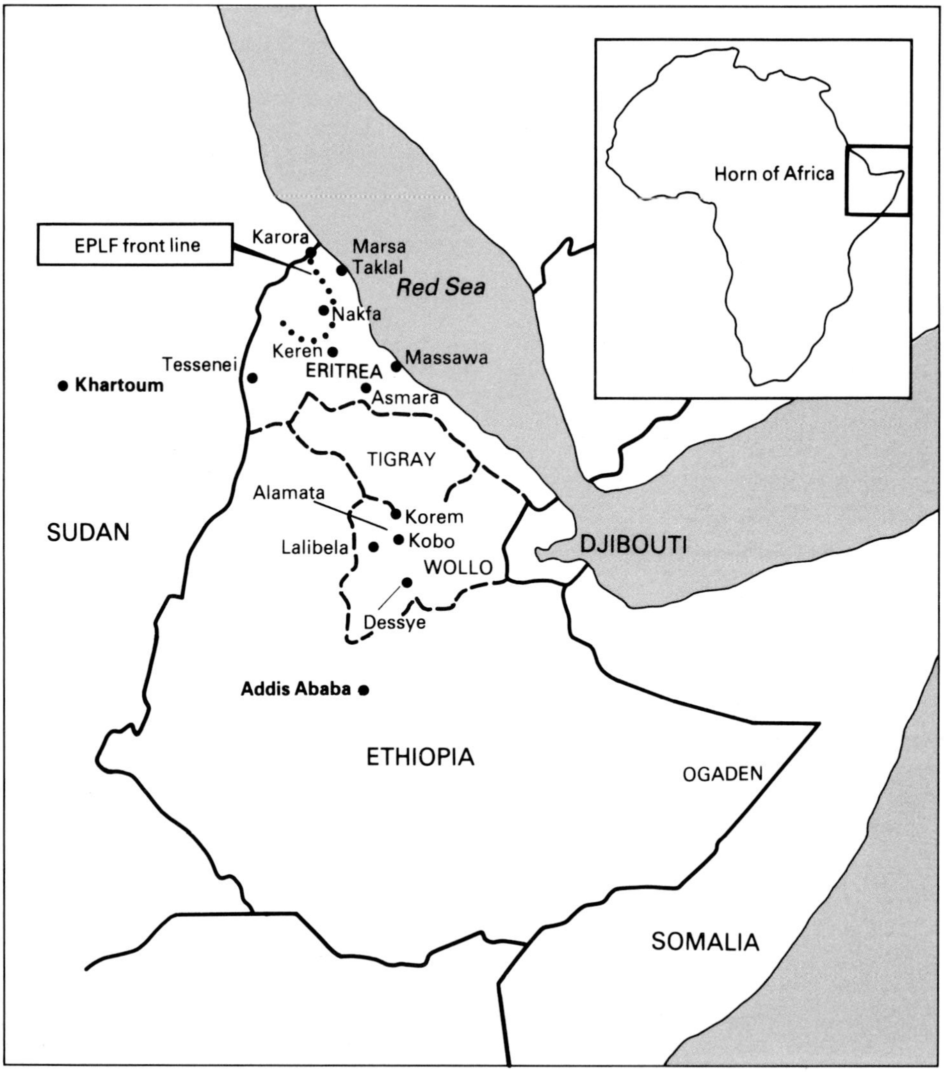

Horn of Africa

War in the Horn of Africa

COMPLEX CONFLICT

Background

This conflict is a complex of inter-related wars involving Ethiopia, Somalia, Eritrea, Tigray and internal minority guerrilla groups. The Soviet Union and Cuba are heavily involved, while the US is less committed. The Russians want ports on the Horn of Africa from which they can dominate and in time of war cut the oil-tanker routes from the Middle East oil fields. The US needs bases to counter this threat.

The seeds of war were sown in 1952 when Eritrea became part of Ethiopia. In 1964 various armed groups affiliated to the Eritrean Liberation Front (ELF) began to fight for independence. Guerrillas had many successes against government troops in Tigray province as well as Eritrea.

In 1974 a military coup in Ethiopia overthrew the monarchy and Colonel Mengistu became virtual dictator. He continued the war against the ELF, but by the end of 1977 the 40,000 ELF fighters were winning the war against 80,000 Ethiopian troops. Only one city, Asmara, remained to be captured.

The tide of war turned when the Soviet Union, which had been heavily supplying Somalia with arms and equipment for its war against Ethiopia, changed sides and backed Ethiopia instead. With Russian military planning and massive logistical help, the army forced the guerrillas from all the towns, though ELF continued to hold the rural areas. The Ethiopians had massive casualties—22,000 dead and 50,000 wounded.

Simultaneously with the Ethiopian–Eritrean conflict, the Tigrayans of the Tigrayan Peoples' Liberation Front (TPLF), related to the Eritreans linguistically and ethnically, were fighting for autonomy within a federation. Also, in the south of Ethiopia in the Ogaden desert Somali tribes were struggling to unite with their countrymen in Somalia. Many fled to Somalia during the Ogaden War of 1977 when Somalia went to war against Ethiopia.

In 1981-2 Somalia fought against an Ethiopian invasion. Ethiopia was not trying to launch a large-scale war against Somalia, but believed that the time had come to give the *coup de grâce* to what Mengistu and his Russian and Cuban advisers took to be a tottering pro-Western regime. They had under-estimated the ability of Somalia, under the leadership of President Syaad Barre, to resist. The Ethiopian offensive of 1982 prompted the US to grant $50 million in military aid, in return for access to strategic Somalia ports, such as Berbera.

Another six or seven guerrilla organisations fought against the Ethiopian Army, which by 1984 had armed forces totalling 306,000, by far the largest in black Africa. Despite repeated operations, including the much publicised Red Star Offensive of 1982/3, these well-equipped forces could not defeat the irregulars.

For the EPLF, which had supplanted the ELF as the main Eritrean guerrilla group, the battles of Tessenei, January 1984, and Marsa Teklal, March 1984, marked a turning point in the long struggle. The EPLF, using mechanised units, overcame the Ethiopian forces by sheer firepower. At Marsa Teklal they threw into the battle 10 tanks, anti-aircraft guns mounted on vehicles, 50-mm and 76-mm guns and 120-mm howitzers. All these had been captured from the Ethiopian Army.

The Eritreans faced 2 regular brigades, 2 militia brigades, a battalion of T-55 tanks, 2 battalions of heavy (120-mm) artillery, a battery of 130-mm guns and a unit of armoured reconnaissance vehicles. Their tactics were good; they gave the impression that they were retreating when in fact they made a semi-circular movement to cut the Ethiopians' supply line—and also their retreat—70 km south of Marsa Teklal.

In November 1984 EPLF fighters wiped out an entire army unit of 400 troops at Barentu; this was a blow to the Ethiopian Army's morale, which was already low.

The War in 1985

At the beginning of the year the Ethiopian armed forces faced these main guerrilla armies:

ELF:	Eritrean Liberation Front: 6,000.
ELF-PLF:	A union of the ELF and the People's Liberation Forces: 10,000 fighters in all.
PLFRG:	People's Liberation Front Revolutionary Guard; about 5,000.
EPLF:	Eritrean People's Liberation Front; about 12,000.
TPLF:	Tigray People's Liberation Front: 5,000.
WSLF:	Western Somalia Liberation Front: perhaps 4,000.

The Ethiopian government, during 1985, spent half its budget in efforts to crush the guerrillas, with the help of 1,600 Soviet, 3,500 Cuban and 300 East German advisers and technicians.

Throughout 1984-5 Ethiopia, in common with much of north-east Africa, suffered from a disastrous drought in which millions of people died. However, the drought did not stop the war and only rarely did it affect government operations. The Mengistu regime consistently refused proposals by the rebels for cease-fires or safe passage for international relief convoys. UN officials and independent relief organisations reported that the administration was frequently diverting international food aid to its troops and militia and was also using famine relief as a political weapon.

For instance, farmers who reported to government food centres were beaten and turned back empty-handed for not having peasant association membership cards; all farmers in government-controlled areas were organised into peasant associations that double as local militias. When the men came back with a card they were given food.

There is little doubt that both the EPLF and TPLF were also intent on obtaining as much political advantage as possible from the drought crisis.

EPLF held more than 9,000 Ethiopian prisoners at the end of 1985. Until 1977 the Front had released most of the prisoners it took, but many were then gaoled by the Ethiopian government or simply vanished. Colonel Mengistu could not admit that any of his soldiers had been captured, since that would mean admitting that EPLF controlled part of Eritrea. Captured soldiers released by EPLF and returning to their homes were an embarrassment. By keeping these men, says the EPLF, it is saving their lives.

Whatever the Ethiopian government refuses to admit, EPLF in 1986 controlled northern Eritrea from the Red Sea to the Sudanese border, including most of the Sahel and Barka provinces. EPLF's forward defences consist of a 400-km line of trenches; the battlefront looks something like that of World War I with just 100 m between the opposing trenches in some places.

In areas where the guerrillas are active, the government troops can move only in heavily armed convoys or they are restricted to towns guarded by military garrisons or local militia. The Ethiopian Air Force, unable to find military targets, often attacks villages. The planes, Mi-21s and helicopter gunships, usually appear on market days or when food is being distributed. Civilian casaulties have been heavy.

The Eritreans work at night, by long practice. To make raids by the Ethiopian planes fruitless, there is little visible activity during the day. Life begins after sunset, when generators are switched on and work goes on under electric light until midnight. Nearly all travel is by night.

EPLF fighters are masters of camouflage and not one of their tanks has been attacked from the air. The Ethiopian reconnaissance planes have also not spotted the plastic factory which makes sandals for the fighters or the garages which repair and maintain captured Soviet trucks. EPLF's main hospital stretches for 5 km along a valley and is half-buried in the gound for better camouflage. Highly developed, the hospital has three operating rooms and produces its own intravenous serum. The guerrillas run their own schools, literacy programmes and relief societies. They hold several towns in the extreme north-western part of Eritrea near the western border with Sudan, and they control numerous rural settlements and villages.

Many Eritrean fighters are Christian highlanders, but there are also Muslims. Men and women mix freely in EPLF, which is remarkable in the predominantly Muslim Horn of Africa.

While the greater part of all types of EPLF arms and equipment has been captured from the army, and is therefore of Soviet-bloc manufacture, the Eritreans receive "gift" supplies from Kuwait, United Arab Emirates and Somalia. This is of mixed American, French, British, Italian, German and Soviet origin. The EPLF fighters have become accustomed to handling weapons of diverse mechanisms.

The Tigrayan fighters of EPLF have also captured many hundreds of Ethiopian guns and mortars and some tanks. Nearly all their small arms are Russian-made—also captured—but they have French and American weapons as well. For the TPLF the major threat is the Ethiopian attempt to resettle 2 million people from the north on more fertile land. The motive is military, not humanitarian; Mengistu wants to depopulate Tigray, which has a population of 4 million, so that village support for the guerrillas becomes non-existent. TPLF guerrillas attack the convoys forcibly evacuating village people and many escape.

During 1985-6 the war between Ethiopia and Somalia merely simmered, but the Somali government is fearful of renewed attack. If the Soviet Union cannot acquire Red Sea and Horn of Africa bases through Ethiopia–Eritrea, the Somalis reason, it may try to reach the coast by breaking through Somalia. For this reason Somalia maintains an armed force it cannot afford—62,500 men out of a population of 6 million. It is one of the largest in Africa, but it is demoralised by poor conditions and low pay. Somali soldiers often cross into Kenya, seeking refuge. Unlike the Soviet Union, which before 1977 lavished arms and equipment on Somalia, the US is unwilling to give more than is necessary for training and basic defence. This is partly because long-term considerations suggest that Washington, like Moscow, considers Ethiopia far more important than Somalia and does not want to jeopardise the possibility of improving relations with Addis Ababa.

Anti-government guerrilla groups exist within Somalia. The main ones are the Somali Democratic Salvation Front and the Somali National Movement. Both groups, which are backed by Ethiopia and Libya, oppose the authoritarian, corrupt, and oppressive Barre regime. Captured guerrillas are summarily executed, as are many non-guerrilla dissidents.

During 1985 and 1986 Italy supplied the Somalian armed forces with 100 M-47 tanks, 300 Fiat armoured fighting vehicles, 14 Marchetti aircraft for counter-insurgency missions and light attack, 4 Aeritalia transport aircraft and 4 Piaggio aircraft for transport and reconnaissance. The Italian Prime Minister made Italy's position clear.

> "Somalia is for us a friend, a great friend, and our position in the Ethiopia-Somalia war of attrition is obvious. We think that the Cuban forces and Soviet military advisers should return home. But we don't believe that a superiority of Somalian armed forces could prevent an Ethiopian attack. A negotiated solution is needed to prevent a serious conflict in a vital part of the world."

The Ethiopian-Eritrean-Tigrayan-Somali war is likely to go on for many years, for these reasons:

(i) The Ethiopian government will never give up control of Ethiopia's outlets to the sea on its northern (Eritrean) coast.
(ii) The Eritreans will settle for nothing less than independence and/or a UN-supervised vote on self-determination.
(iii) The EPLF and TPLF know that they can fight and survive indefinitely. Warfare in both its conventional and guerrilla aspects is a way of life for them.
(iv) The Soviet Union, after its tremendous political, economic and military investment in Ethiopia, cannot withdraw—or allow the Ethiopians to make a negotiated settlement.
(v) The Organisation of African Unity will not intervene, and dismisses the conflicts in Eritrea and Tigray as "internal affairs"; the UN has also done little to encourage a peaceful settlement.

During 1986, tensions between Sudan, Ethiopia and Somalia appeared to ease. If this were to end in an alliance, no matter how uneasy it might be, the liberation groups in each of the three countries would be left without bases in the others. Ethiopia is particularly anxious to dissuade the Sudanese from backing the Eritreans. According to the EPLF, they killed, wounded or captured 29,500 Ethiopian troops during 1985. In addition they captured 18 Ethiopian tanks, 138 vehicles, 5 APC and 39 heavy guns.

India–Kashmir War

A FRONTLINE AT 20,000 FEET

Background

On either side of the Jhelum River in Kashmir, the Indian and Pakistani armies have confronted each other across a UN ceasefire line since 1947.

India holds two-thirds of Kashmir and Pakistan the other third. The Indian government believes that Kashmir's accession to India in 1947 was legal and final, and wants the ceasefire line to be recognised as an international boundary. Pakistan wants the future of the predominantly Muslim state, which has a population of 7.5 million, to be determined by a UN-sponsored plebiscite.

At a height of nearly 20,000 ft the opposing armies have been no more than 500 m apart, though the ceasefire line was revised slightly in 1972 following the India–Pakistan war of 1971.

Thousands of ceasefire violations have taken place, but dedicated work by the 40 UN observers over a period of 36 years has nearly always prevented open hostilities. The observers come from Australia, Belgium, Chile, Denmark, Finland, Italy, Norway, Sweden and Uruguay.

The dispute in Kashmir is not confined to its religious and geopolitical importance. It is also an argument over water; the headwaters of the rivers that flow through Pakistani Kashmir originate on the other side of the frontier in India. For Pakistan, there is yet another factor: the Siachin glacier in Kashmir overlooks the 800-mile Karakorah highway which links Pakistan with Peking. China is Pakistan's ally.

In May 1984 the Pakistanis launched a major infantry–artillery attack to dislodge Indian troops from the glacier. An entire company of Pakistani soliders vanished on the Indian side of the dangerous glacier, probably killed by an avalanche.

The War in 1985

Early in the year fierce fighting took place between Indian and Pakistani units; the Pakistanis suffered more than 50 casualties and the Indians 30; most of the Indians died in an avalanche brought on by artillery fire.

In June 1985 further fighting broke out on the glacier, which is 48 miles long. Pakistani B57 bombers made many bombing runs against Indian positions. A number of air duels took place and India claimed to have shot down two Pakistani aircraft.

According to UN observers, casualties on both sides ran into hundreds. The UN officers are not directly involved on the glacier, as it was not part of the original demarcation line. However, because of the length of the ceasefire line—nearly 700 km—and the increasing number of violations, the UN commander in Jammu-

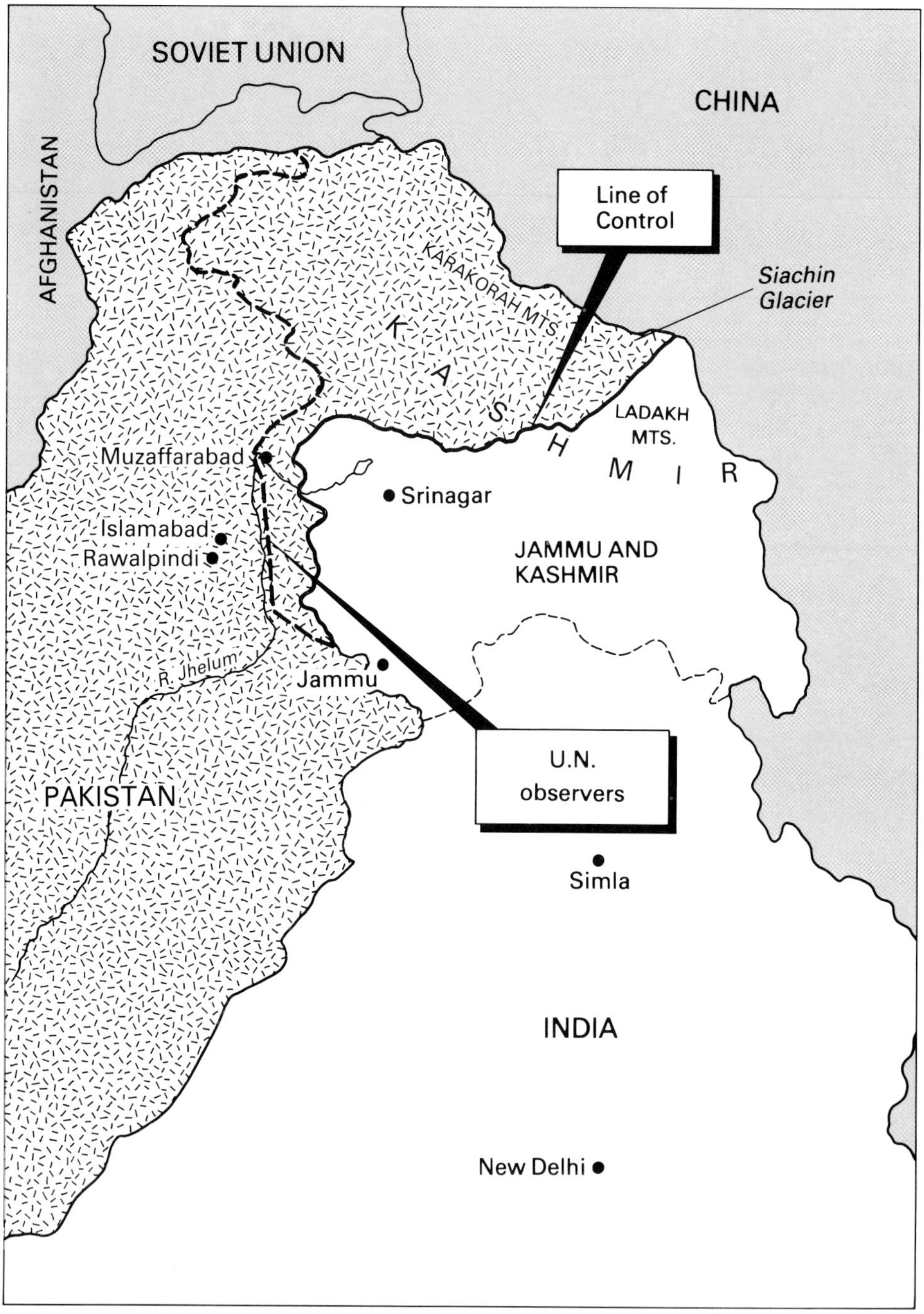

India – Kashmir War

Kashmir asked for another 25 observers. The UN Secretariat would like to withdraw the observer group, but Pakistan, fearful of an Indian invasion, will not agree.

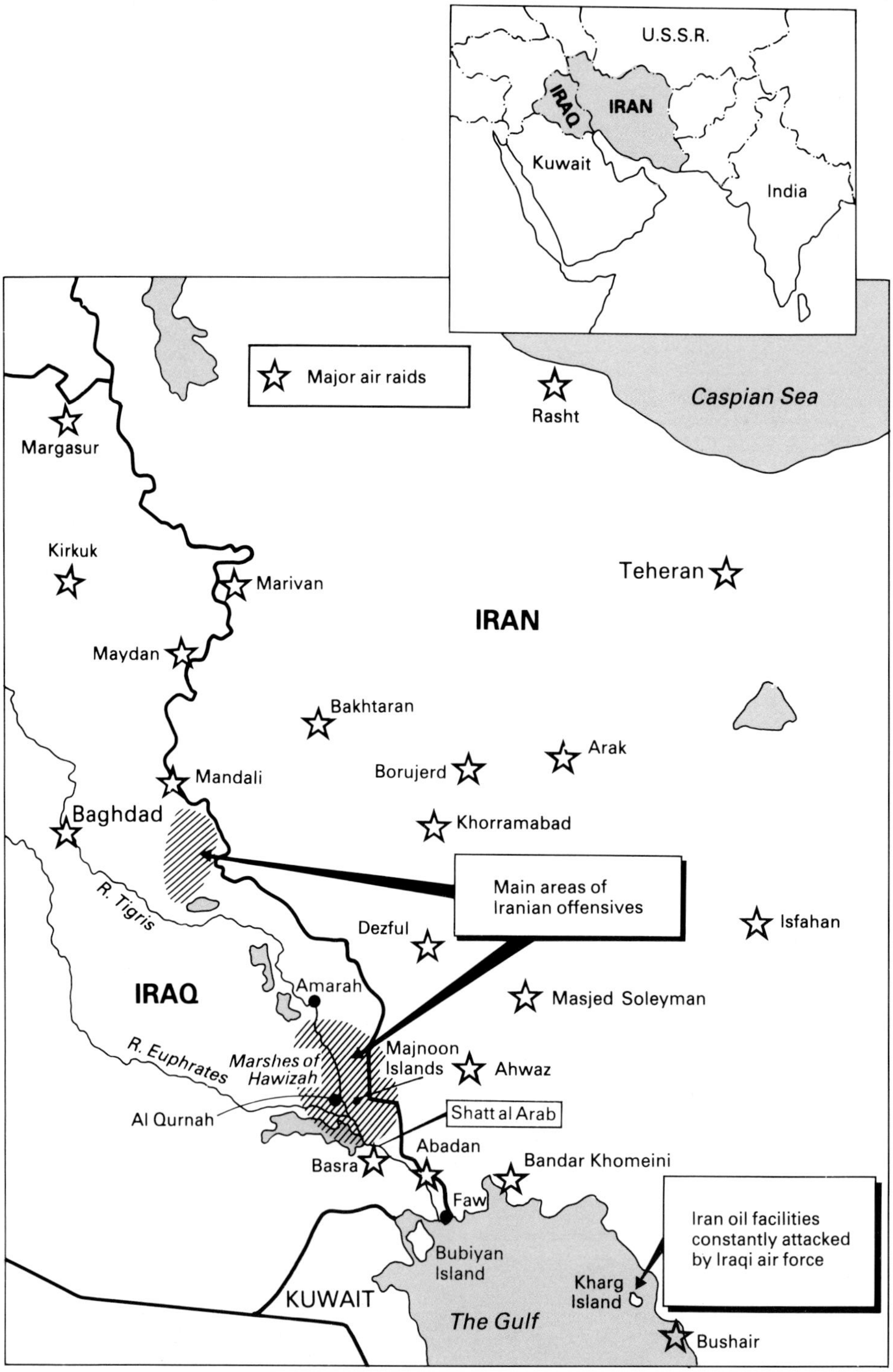

Iran – Iraq War

The Gulf War

MASSIVE CONFLICT

Background

The Iran–Iraq War, generally called the Gulf War, began on 17 September 1980 when the Iraqi president, Saddam Hussein, ordered his army to capture Iranian territory which he claimed really belonged to Iraq.

Iraq is dependent for its life on the Tigris and Euphrates rivers, which come together 120 miles from the Persian Gulf, to form the Shatt-el-Arab. This waterway is Iraq's only outlet to the Gulf through which it exports its oil, the basis of its wealth.

Iraq claimed the whole channel up to the Iranian shore as its territory, while Iran insisted that the Thalweg, a line running down the middle of the Shatt-el-Arab, was the official border.

Several treaties were signed over this dispute, the outcome usually based on the relative strengths of the two countries. The last, achieved in 1975 through Algerian mediation, was to lessen the tension resulting from the Iranian seizure of the Hormuz Islands in 1971, as the British pulled out of the Gulf. An arrangement was reached, under which the Shah of Iran consented to withdraw his support for the drawn-out Kurdish revolt against the Iraqi government; in return, Iraq was forced to acknowledge Iranian sovereignty up to the middle of the Shatt-el-Arab. For Iraq, the treaty was humiliating.

Another lasting tension was based on religious problems. Iraq has a 55% Shi'a Muslim majority while Sunni Muslims make up only 35%. The Iraqi ruling caste, however, is not only Sunni but leftist-orientated and cares little for religion, the very basis for the Islamic Revolution brought about by Ayatollah Khomeini of Iran.

Khomeini, persecuted by the Shah in the 1960s, found sanctuary among the Shi'as of Iraq, but the rulers of Iraq's Ba'ath Party expelled him. He remained in hiding in France until his triumphant return to Iran following the overthrow of the Shah's regime in 1979.

As soon as the Shah fell, the Iraqis claimed that the "infamous" treaty over the Shatt-el-Arab must be revised. Their demands became more forceful as they watched the once powerful Iranian Imperial Army disintegrate, most of its commanders executed by Khomeini's Islamic courts. Border clashes and local skirmishes were regular occurrences some months before the "real war".

In the Iranian oil-rich region of Khuzestan (Arabistan to the Iraqis), where the majority of the population is Sunni, riots incited by Iraqi Intelligence officers broke out over labour disputes. In the north, a new Kurdish rebellion—this time within the borders of Iran—caused the Khomeini Islamic government severe troubles. This, too, was backed by the Iraqis, despite the danger of inciting Iraq's own Kurdish minorities, still unsatisfied with their position in Iraq.

Beyond the immediate aim of regaining territory, Saddam Hussein, President of Iraq since 1979, had much bigger objectives:

To make Baghdad the power centre in the Arab Middle East.
To replace non-Arab Iran as the dominant Gulf nation.
To undercut any counter-effort by revolutionary Shi'a Islam in Iran to subvert Iraq through its large Shi'a Muslim community.

As Hussein and his general staff planned their campaign, he had every reason to be confident. The Iranian Army lacked first-class leadership, because the senior and more intelligent Iranian generals had been executed by the revolutionary regime of Ayatollah Khomeini. In addition, the Intelligence staff told Hussein that the Iranian armed forces could not long sustain a war because they lacked spare parts for their tanks, armoured personnel carriers and aircraft. Much of the equipment was American-made and as the US had cut all ties with Iran, the Iranian forces would soon be crippled.

While the Iranian armed forces were weak and badly led, the Iraqi armed forces were at their peak in equipment, morale and training. Against the Iranian Revolutionary Guards, led by religious mullahs with little military experience, the Iraqis could muster 12 complete mechanised divisions, equipped with the latest Soviet Union material. Since the last combat involvement of Iraqi forces, during the later stages of the Yom Kippur War (1973), much had been done to re-equip the armed forces and organise new formations. The air force had been given special consideration and priority, bringing its first-line aircraft to over 300, most of these second-generation jets of highest quality.

The ground facing the invasion axes caused the Iraqis no anxiety. The border area was well across the river and no major obstacles faced the first stage of an attack. For an army so well equipped with Soviet river-crossing material, the Kharkeh and Kharun rivers posed no immediate problems; the Iraqi command assumed that crossing sites were probably lightly defended against a fast-moving armour and possibly airborne attack.

Iraqi Intelligence sources reported that of the former Iranian order of battle in Khuzestan, which included two divisions based in Ahwaz, Dezful and Abadan, only battalion-sized formations would be encountered, and even these were ill equipped and of low morale.

The area had formerly been controlled by the regional 1st Corps HQ at Khermanshah farther north—now non-existent from a military standpoint; control of operations would be directly from Teheran, with all the handicaps involved. As for tank units, only a handful of company-sized units were operative. Most of the armour was inactive in camps where it was poorly maintained and technically neglected.

The Iraqi planners' only uncertainty was the operational value of the Iranian Air Force. One of the few facts known to Iraq was that most of the former Iranian commanders and many operational pilots were either executed or abroad. However, the Iranian Air Force had shown up during the local riots and following the US operation in the Salt Desert in an attempt to rescue American hostages. This proved that at least some of its combat aircraft were still operational. The danger from an eventual Iranian counterstroke on the armour massing to cross in Khuzestan worried the Iraqi planners considerably. To eliminate this danger they decided on a massive pre-emptive air strike on the Iranian air bases in an effort similar to that employed by Israel during the Six-Day War of June 1967.

On the morning of 22 September 1980, the Iraqi Air Force raided 10 major Iranian air bases. Formations of Iraqi Mi-23s and Mi-21s swooping low over the Iranian countryside attacked Mehrabad, Doshen-Tappen at Teheran, Tabriz, Khermanshah, Ahwaz, Dezful, Urimiya, Hamadan, Sanadai and Abadan. Also attacked were two early warning stations at Dehloran and Neft-E-Shal. Iraq completely surprised the Iranian defences, but the results were negligible. Soon after the Iraqi jets disappeared over the hills to the west, Iranian Phantoms took off from the same bases for a retaliatory attack on Iraqi targets. The Iraqi air strike failed because most of the Iranian aircraft were in protective hangars, some of them Israeli-constructed during the Shah's regime. Although runway-busting bombs were dropped, they did not put the large airfields out of commission. Emerging from their shelters, the Iranian Phantoms, with bomb loads three times larger than the Iraqi jets could carry, smashed strategically important targets close to the main population centres, causing great destruction.

Six Iraqi divisions crossed the frontier at four points simultaneously. In the north, a mechanised division overwhelmed the Iranian border garrison at Qasr-Shirin on the Khermanshah–Hamadan road, the major access route to Teheran. This attack, although impressive, was only a diversion to the main attack launched by five well-supported armoured and mechanised divisions which smashed into Khuzestan on two different axes. One division, crossing the Shatt-el-Arab bridge at Basra, made for Khorramshahr along the main road; a second division headed for Ahu on the Ahwaz road, cutting off the battle zone and securing the Kharun River. A third division entered Iran on the El-Amarah road towards Susangard with Ahwaz, the major military base in Khuzestan, as its objective. This thrust was supported by another division-sized force making for Dehloran and Dezful air base farther to the north, to cut off reinforcements coming down the Zagros Passes from Khermanshah and Teheran. A sixth division was held in immediate reserve at Basra to push through Khorramshahr towards Abadan to capture the vital oilfields.

The ground attack was carefully planned and reinforced with heavy armour and artillery to ensure rapid execution with overwhelming power. Again, it was believed nothing the Iranians could do would stop it. In hours, the region was engulfed in fire and smoke as massive artillery concentrations pounded oil storage tanks, reservoirs and pipelines.

The Iraqis' timing was ideal; they achieved tactical surprise and the Iranian military command was in crisis. Military analysts, politicians and journalists from all over the world predicted that the highly professional and well-equipped Iraqi Army would win the war in a matter of weeks.

"Given the limited military capability of the combatants, the war does not appear likely to be a long one", *Time Magazine* commented on 6 October 1980.

The Iraqi attack was likened to a blitz and to "an Israeli-type" assault. In fact, the Iraqis had closely studied Israeli tactics used in the 1967 (Six-Day) War and in the 1973 (Yom Kippur) War.

The Iraqi commanders achieved their aim of capturing the Shatt-el-Arab and their troops seized a 30-mile strip of Iranian territory, but could not break the Iranian line. The Iranian leadership did not, as Hussein expected, accept defeat and propose peace.

Iranian heavy artillery—155-mm American-made guns shelling Iraqi fortifications on the southern front.

Iraqi troops firing on Iranian positions on the northern sector of the front. They are using the Soviet-made RPG Mk IIs. The hollow-charge rocket penetrates 40 cm of armour at a range of up to 800 m.

The reasons for the stalemate and the war of attrition which developed are the most interesting aspects of the Gulf War.

For the first year the conflict was limited to artillery duels, air attacks and patrol actions. In 1981, as more Iranian commanders returned to their posts, after "disloyal" personnel were pardoned for the sake of the country, morale rose; the Iraqi forces were halted on the Kharun River and the counter-attack drove the invaders beyond their borders. This intensive phase of the war did not last long because of the limited stocks of war material.

Each of the belligerents found supporters. Fearing the spread of Khomeini's Islamic revolution, Egypt, Saudi Arabia and Jordan backed Iraq. Syria and Libya backed Iran, mainly because they were opposed to Iraq's allies.

In July 1982 Iran launched *Operation Ramadan* against Iraq, and for the first time in the war fighting took place on Iraqi soil. In one of the biggest land battles since 1945, Iran used wave after wave of Revolutionary Guards (the Pasadan), more popularly known as Soldiers of Allah or holy war warriors. Aged from only 9 years of age to more than 50, these eager but relatively untrained soldiers swept over minefields and fortifications to clear safe paths for the tanks. Their casualties were immense, but in places they swamped the Iraqi defenders and recovered some territory before the assault petered out.

By the end of that year Iraq was massively re-equipped with Soviet material, endless rows of T-55 and T-62 tanks, "Stalin organs" (rocket launchers), Mi-24 helicopter gunships and thousands of tank transporters; eventually each tank had its own transporter. (During the 1973 Arab-Israel War most Iraqi tanks were sent to the distant front on their own tracks along rough roads and were wrecks by the time they arrived.) Iraq also prepared a Soviet-type three-line defence, entrenched with obstacles, minefields and fortified positions. The Combat Engineer Corps was probably the most efficient part of the Iraqi Army. From the beginning it was intensively used for assault bridging of water obstacles, for laying minefields, building strategic bridges and for preparing new defence lines and fortifications.

On 6 February 1983 the Iranians attacked with 200,000 troops, most of them Revolutionary Guards, along a front of 40 km, in the mud and desert of Ammara, south-east of Baghdad. The six-division thrust was decisive enough to break through. Backed by air, tank and artillery support, most of the Iranians' last reserves were committed.

With its last reserves going to the battle, Iraq used massive air support, with over 200 sorties a day and intensive use of attack helicopters. Over 6,000 Iranians were killed that day achieving only minute gains.

In April 1983 the Iranians struck again in repeated attacks along the axis from Mandali to Baghdad, in the north-central sector. The Iraqis stopped these moves by using their mechanised and infantry divisions in rigid, heavy defensive lines.

Once again casualties were heavy, especially among the Iranians. By the end of 1983 120,000 had been killed and 30,000 taken prisoner. Iraq had lost half this number.

Despite its heavy losses and the enormous cost of waging the war, Iran re-emerged at the end of 1983 as a major third-world economic power. It was exporting much more oil than Iraq, and with three times Iraq's population it had the better chance of winning a war of attrition. In readiness for the final blow, much of Iran's regular infantry force was still kept out of battle.

France, seeking to protect a £10,000 million investment in Iraq, began a daily airlift of military supplies and "leased" to Iraq five Super Etendard warplanes equipped with Exocet missiles. The year 1984 promised to be even more violent than the first three years of the war.

The Opposing Forces in 1980

Iraq

Army: 200,000, 100,000 reserves

Armoured divisions	4
Independent armoured brigades	1
Mechanised divisions	2
Infantry divisions	6
Independent infantry brigades	1
Special Forces brigade	1

Iraq's armoured divisions are exceptionally large. The 300-500 tanks of an Iraqi division outnumber by about 80 those of armoured divisions in most other Middle East armies.

Equipment:

Tanks	2200 (T-54, T-55, T-62, T-72, T-59, PT-76)
APC	3000
Other AFVs	Mainly BRDM-1/2, FUG-70 and ERC-90
Artillery	800 guns (towed only) of 85-mm, 100-mm, 120-mm, 130-mm calibre.
Air defence	1200 guns, ranging from 23-mm to 130-mm as well as SAMs.
Anti-tank	107-mm, 75-mm, 85-mm, 100-mm, 105-mm self-propelled. ATGW, SAGGER, SWINGFIRE, HOT, Milan.

Navy: 4,500 men.

Frigates	Lupo class,	4
Minesweepers	T-52 (2)	Yevgenia (3)
Missile boats	Osa (12)	Wadi, 6 under construction, Italy
Gunboats	P6 (12)	
Patrol craft	10	
LCTs	4	

Iraq's small navy is constrained by the country's narrow, 58-km coastline. It consists largely of Soviet-made missile boats. Because of Iranian naval superiority, the Iraqi Navy can barely operate in the northern reaches of the Persian Gulf. When ships ordered from Italy were received in 1985 the navy grew in power and status.

Air Force: 18,000 men, 530 aircraft

Interceptors	10 Mi-25A
Fighters	75 Mi-23, 150 Mi-21, (35 Mirage F-1EQ)
Missiles	AA6, AA2, Magic 1, R530
Bombers	8 Tu-22, 12 Tu-16, 4 Il-28
Helicopters:—	
AT	12 Mi-24, 15 SA-342 Gazelle
Assault	100 Mi-8, 14 Mi-14, 10 Puma

Also trainers, heavy transport aircraft and liaison planes.

The Iraqi Air Force has traditionally been equipped with Soviet-made aircraft, but Mirage F-1s have been acquired from France as part of an effort to diversify arms sources. During 1984 Iraq built up one of the largest helicopter fleets in the Middle East.

Iran

The figures given here applied when Khomeini's Iranian Revolution began. The numbers of tanks, guns and aircraft which remained battleworthy on the outbreak of the war with Iraq were considerably less.

Army: 300,000 men with reserves of 300,000

Armoured divisions	3
Infantry divisions	3
Independent brigades	4
Anti-aircraft missile battalions	4
Army Aviation Command	About 100 aircraft.

Equipment:

875 Chieftain tanks	(designated FV-4030), 400 M-47/48
	460 M-60A1, 250 Scorpions, plus Fox and Ferret Scout Cars
APC	2000, mostly M-113s and BTR 40/50/60/152s
Artillery	710 guns, mostly 155-mm self-propelled.
ATGW	SS11, SS12, TOW
Air defence:	1800 guns; 100 ZSU-23-4, ZSU-57-2, SAMs and HAWKs
Helicopters	(separate from the Air Force helicopters)
	621

Navy: 28,000 men

Destroyers	3
Frigates	4
Corvettes	4
Large patrol craft	7
Fast guided missile boats	5
Minesweepers	5
Landing ships	2
Support ships	2
Other craft	14

Air Force: 100,000 men; 459 combat aircraft.

The main elements of the air force were 188 Phantom F-4s, 166 F-5s, 77 advanced F-14 interceptors and 163 helicopters. The air force was well equipped with missiles, notably Phoenix, Sidewinder, Sparrow, Maverick and Condor.

Taken together, the Iranian armed forces were formidable until the Revolution. By the time the war broke out, only 8 of the F-14 interceptors were airworthy; probably half the combat planes could not be used because of lack of parts or maintenance or because of sabotage. One-third of the Chieftain tanks—the Shir-Iran ("Lion of Persia") version—were no longer serviceable. The army's manpower was down to 180,000 as a result of purges and desertions. At least 250 generals had been replaced by inexperienced officers or military-minded mullahs.

The War in 1984-5

During 1984 the war showed the chaotic nature of third-world conflicts. There had been a propensity to believe that such conflicts would resemble the 1973 Arab-Israel (Yom Kippur) War, with each side using advanced weapons in a conventional manner to defeat the other's army on the battlefield in a fairly organised manner. But neither Iraq nor Iran have used their modern equipment with any imagination or proficiency. Their armies have shown little co-ordination, and some units in the field have been left to fight largely on their own. Huge amounts of sophisticated materiel were left unused when a massive, modern assault could have won the war for either side.

Observers of the continued fighting between the Iraqi armed forces and the Iraqi-Kurdish rebels have not been impressed by the quality of Iraq's soldiers and commanders, nor by unit combat standards. The Iraqi performance on the Syrian front during the 1973 war against Israel was also not brilliant. Some observers had been convinced that the vast expansion following 1973 was accompanied by an improvement in combat standards. This conviction has not been borne out in the Gulf War. For instance, senior commanders in the field refer too many "difficult" decisions back to Baghdad.

The first major event of the Gulf War in 1984 was an Iranian offensive, code-named Dawn V, which began on 22 February. In a surprise attack the Iranians hit the junction point between the 3rd and 4th Iraqi Army Corps and had a temporary success.

On 27 February the offensive continued with a massive assault of 500,000 men, mostly Revolutionary Guards and young volunteers, across the marshlands north of Basra. By shallow boat and on foot they moved to within 4 miles of the Basra–Baghdad highway without seriously threatening to cut it.

The next phase of this operation led to a big battle in the flat desert, just north-east of Basra, on 29 February and 1 March. The Iranians had inadequate air and armoured support and the Iraqis attacked them with tanks, mortars and helicopter gunships. More than 19,000 Iranians were killed and the force was driven back. The sides settled into defensive lines about a mile apart.

While this land fighting was going on, Iraqi ships and planes were attacking Kharg Island, the Iranian oil terminal.

In the second week of March heavy fighting occurred in the reed-filled marshes and shallow lakes of the Hawizah Marshes, mostly in Iran, just east of Qurnah, near the confluence of the Tigris and Euphrates rivers. This wilderness was held by five

Iranian brigades—15,000 men. In Soviet and French-made helicopter gunships flying at low level, Iraqi special forces searched out the enemy units and inflicted heavy casualties in what the Iraqis call the Battle of Majnoon.

Without the equipment to force passages through the Iraqi minefields, and short of tanks, the Iranian command again resorted to the "human wave" tactic. In March an East European journalist saw tens of thousands of children, roped together in groups of about 20 to prevent the faint-hearted from deserting, make such an attack. They hurled themselves onto barbed ware and marched into the minefields in the face of machine-gun fire to clear the way for Iranian tanks. Again there was little progress, in spite of the sacrifice of children.

At the Battle of Gzaeil that month, the Iranians for the first time used a regular army unit, 92nd Armoured Division, but it too achieved little.

In a period of 4 weeks in February-March the Iraqis killed 40,000 Iranians and lost 9,000 of their own men, but even this was an unacceptable ratio and in February the Iraqi command ordered the use of mustard gas.

For 18 months Iraq had been producing mustard gas, at Samawah, and nerve gas, at Rutbah. Both weapons are outlawed by the Geneva Protocol of 1925, to which Iraq is a signatory, but the government was desperate for a way to counter the overwhelming Iranian manpower. Gas was an "area effective" weapon, useful for dealing with slow-moving infantry attacks and with many thousands of troops concentrated in a relatively small area.

Employed on the southern front, it caused at least 1,000 Iranian deaths before international public opinion, political pressure from Iraq's allies and an Iranian threat to use chemical weapons in retaliation caused the Iraqis to abandon its use.

In March 1984 it became clear, for the first time, that the US was no longer so vulnerable to a possible oil cut-off from the Persian Gulf. That month the Reagan administration announced that in a crisis the US would at once turn to its 400-million barrel strategic petroleum reserve. In effect, this told Iran that its threat to close the Straits of Hormuz was no longer a serious threat to the US. The underlying purpose of the reserve, which is constantly growing, has always been to help prevent the US—and indirectly its European allies and Japan—from ever again being held hostage to the dictates of Middle East oil-producing nations. That happened during the embargo by the OPEC countries in 1973. Use of the oil from the "war reserve"—as American strategic planners call it—could buy the US up to 45 days of breathing space.

Iran abandoned its plan to close the straits, but Iraq embarked on a "tanker offensive"—missile attacks on Panamanian, Greek, British, Saudi and American oil tankers in an effort to draw the superpowers, the US particularly, into the field of conflict and to apply further pressure on Iran. In both aims Iraq was successful. US warships stepped up patrols and convoys in the Gulf. Saudi Arabia was supplied with American Stinger missiles and Iraq received regular intelligence from the Saudis' American-supplied AWACs early-warning planes.

Reverting once again to infantry action, the Iran Command turned to the hilly central and northern areas of the border where they have an advantage over the Iraqis, who find it harder here to make proper use of their superiority in aircraft and armoured vehicles than in the flat land farther south.

On 18 October the Iranians attacked on a 12-mile front in the hills overlooking the Tigris plain in an attempt to regain land which the Iraqis took at the beginning of the war. They pushed the Iraqis back over the border, but the Iraqis counter-attacked on October 20 and after five assaults they re-took some lost ground. About 4,000 men, Iraqis and Iranians, died.

Throughout 1984 both sides received fresh materiel from abroad, but the greater proportion by far went to Iraq. At the end of year the army had 5,000 tanks—twice as many as 12 months previously and five times as many as the Iranians had. Their combat aircraft outnumbered those of the Iranians about 8 to 1 and their heavy guns by 3 to 1. The Soviet Union supplied the largest amount of materiel, but in 4 years France sold to Iraq arms worth £4 billion and Saudi Arabia paid the bills. Egypt made a gift of 300 Soviet tanks. The Iraqi Air Force had lost 84 planes to the end of 1984, but the losses were more than made up by Soviet replacements.

The Sovietisation of Iraqi armament in 1984 coincided with a transformation inside the army, involving the gradual replacement of British-trained officers and generals by a new generation of military men from the Ba'ath Party. The party now controls the entire army through its "political organisation officers" who are the Ba'ath's political commissars in the army.

During 1984 Iran received substantial military aid from Argentina and North Korea; the Koreans sent £950 million in weapons, China supplied about 100 F-6 fighter planes and South Africa 300 155-mm guns. Switzerland delivered 8 PC-7 trainer aircraft in 1984. Syria and Libya have supplied 380 main battle tanks, 100 of them during 1984.

Towards the end of the year General Qassemali Zahir Nejad was appointed as Ayatollah Khomeini's second representative on the Supreme Defence Council; Nejad is a more imaginative and innovative general, and his appointment indicated new thinking about strategy and tactics. The new strategy follows these lines:

To threaten Iraq in the south.

To make major repeated attacks in the north directly west towards Baghdad, 150 miles away. Each attack would take some ground until the Iraqis rushed reinforcements to the sector, when the Iranians would stop attacking and again threaten the south. The regular army would be used in the Baghdad push.

To wage a general war of attrition, since time and numbers were on Iran's side.

The fall of Baghdad, Iran's generals and ayatollahs agreed, would mean the fall of President Hussein. But a successful invasion of Iraq would also probably cost Iran the support of Syria. The Syrian leaders would lose all Arab-world sympathy if they backed an Iranian invasion of an Arab country.

Throughout 1984 the war was increasingly unpopular with Iraqis and there were many military deserters. When caught they were shot in batches of 10, in front of their comrades and families, who must then pay funeral expenses.

During 1985 fewer executions occurred, because there was less danger at the front and therefore fewer desertions. Both sides refrained from making massive frontal attacks and devoted the greater part of their military energy to improving their defences. For instance, Iraq completed a great lake and a network of canals around Basra to protect the city. The project is believed to have cost at least a billion dollars. Continuity satellite pictures show that gun and rocket emplacements have followed the canal-digging. Some Iraqis have said privately that in future military operations even more Iranians will perish in the water defences than have already drowned in the lakes and swamps around Shatt al-Arab.

Some geo-strategists speculate that the new system could even be offensive rather than defensive, in that the Iraqis plan to drain out the water from Iran's Khuzistan province. Certain of these rivers, notably the Karun, as well as the canal system which they supply, are Iran's barrier against Iraq. Should they dry up, Iran could be extremely vulnerable to attack.

For their part, the Iranians used tens of thousands of men during 1985 to dig tank traps and to build heavily defended gun positions covering natural tank routes. On the central and northern fronts they have defences in depth so that ground can be traded for time.

Patrol action was intense throughout 1985. The Iranians seemed intent on capturing prisoners for interrogation, while the Iraqis' object was to demoralise the Iranians by slaughtering the garrisons of captured posts.

For the Iraqis, the principal targets of 1985 were the oil installations of Iran's Kharg Island and tankers using its facilities. During 1985 at least 5 tankers were destroyed and another 30 were damaged. By reducing Iranian exports of oil, the Iraqis hope to cripple Iran's economy; also, they hope to provoke superpower intervention to bring an end to the war.

The war is one of the most serious and destructive since 1945. That it was caused entirely by local and regional rivalries is an indication of future troubles among third-world countries. The Gulf War has shown that individual smaller countries are often sufficiently rich and well armed to be able to act on their own regardless of or in spite of the preferences of the great powers.

Because for long periods this war has been one of attrition, it has been likened to World War I. The only real similarity is in the large numbers of casualties. Continuous trenches have not been used to the same extent as along the Western Front of World War I and the two armies have been not mere yards apart but several miles. Western commentators probably likened the Gulf War to World War I because neither side made any more significant progress than did the Allies or the Central Powers between 1914-18.

Iraq always feared that the West would favour Iran because of its size, population, market potential and prime strategic location as the only state barring Soviet entry to the Gulf. In fact, the West has generally favoured Iraq as the lesser of two evils—a victory for Iran would be more destabilising and dangerous than a victory for Iraq. In November 1984 the US resumed diplomatic relations with Iraq after a break of 17 years. The Stockholm International Peace Research Institute claims that both the Soviet Union and the United States are deliberately refuelling the war by supplying both sides and seeing to it, directly or through surrogates, that neither side can win. An exhausted Iran *and* Iraq would suit both superpowers.

No matter how it turns out, the war continues to have the potential to cause serious problems to the West. If Iran wins, the West faces the possibility of a fundamentalist empire hostile to Western interests stretching from the Syrian and Jordanian borders across to the Soviet Union. If Iran loses, there is the danger of its slipping into anarchy, with the Soviet Union coming in to pick up the spoils.

Various people and groups have made attempts to end the war—the League of Islamic States, the UN Secretary General, the Japanese Government. The Iraqis want an "honourable" peace and have attacked Western and Arab shipping in the

Gulf in the hope that a peace may be imposed from outside. The Iranians have never considered peace negotiations.

After more than 5 years of fighting, Iraq held 270 square miles of Iranian territory while Iran claimed to have captured 386 square miles of Iraqi territory. Iranian dead totalled at least 170,000 and possibly 250,000; the Iraqis held 41,000 Iranian prisoners. About 66,000 Iraqis had died and 11,000 were prisoners. The war was costing Iraq £1 billion a month and Iran about half that amount.

At the end of 1985 there was still the danger that the war could spill over into Kuwait and Saudi Arabia, both of which have large Shi'a minorities who consider the Ayatollah Khomeini to be their spiritual leader.

At the end of the fifth year of the war Iran was in the midst of a logistics crisis—where to get the weapons and munitions to carry on the war. Iraqi air attacks on Iranian oil installations had reduced the flow of export oil to an eighth of its capacity, so that the country was short of money to pay for weapons.

Britain permitted the export to Iran of parts for British-made tanks, but the American arms embargo was highly damaging to Iran's prospects. However, Iranian agents managed during 1985 to subvert American nationals who smuggled to Iran, through Britain, much equipment stolen from US navy stores. The thefts include missiles and parts for fighter aircraft. In addition, China was supplying more material than ever before, including Scud long-range missiles.

Israel is Iran's main supplier—or at least its most reliable partner. During 1985 Israel provided arms and equipment worth $800 million. Ayatollah Khomeini and the other ayatollahs have declared their intention to destroy Israel and there is no love lost between the two countries, but it is in Israel's interests for Iran to win its war against Iraq; it would be even better for Israel if the war were to continue for several more years. Iraq is a greater potential enemy to Israel than is Iran.

Israel is now Iran's main supplier of artillery gun barrels and artillery ammunition; a prodigious amount is fired off each week in long-range gun duels. Anti-tank missiles are another major export, principally the American-made TOW missile which Israel defence industries upgrade before re-export. As Iraq has a vast tank fleet, the TOWs are a vital part of Iranian defence.

Libya, with large stocks of weapons and missiles, agreed in the summer of 1985 to sell Soviet Scud missiles to Iran and Iranian emissaries are constantly visiting Libya to plead for greater help with materiel.

At the end of 1985 foreign analysts were convinced that neither side could ever gain a decisive advantage over the other in overall strength.

The Propaganda Campaign

The propaganda emanating from both sides in the conflict is interesting because of its Islamic nature, abusive presentation and because of the intensive efforts by the propagandists to inspire their own peoples.

Teheran radio presents the conflict as "a war between Islam and heresy"—that is, between Ayatollah Khomeini's fiery type of Shi'a Islam and the rule of Iraq's secular President Hussein. Khomeini calls Hussein "the epitome of atheist filth". In the middle of 1984 Iran took to calling its Iraqi foe "the Zionist enemy"—the Arab nationalist term for the Israelis and therefore a deadly insult.

Teheran consistently calls for revolution across the border. In a broadcast in October 1984 it broadcast an "amnesty" for those Iraqis who aided the Iranians so that an Islamic government could be established in Baghad. Repeated broadcasts call on Iraqi troops to surrender and instruct them what to chant so that they will not be fired on—"Allah is great. Victory to Islam and death to Saddam".

Ayatollah Khomeini in nationally broadcast remarks to government news media employees lamented that Iraqi propaganda had successfully hidden the fact that Saddam is so desperate that "the poor man has come down in the world so much that he is going around borrowing money and begging others to come and restore the peace".

Internally, the Iranian leaders and their propagandists present the conflict as a holy war not only to defeat Iraq and its allies but eventually to conquer Israel and Lebanon. The mullahs had no difficulty in inducing parents to allow their teenage sons to enlist because they promised "martyrdom and therefore paradise" to all those who die in the war. Iran radio has admitted that entire classes of schoolboys have "become martyrs". Not until late 1984, after tens of thousands of boys had been killed in suicide attacks, did parents begin to protest about juvenile call-ups.

The Department of National Instruction issued a battlefront primer of 193 pages entitled *Book of Souvenirs for the Front and for the War.* The bottom corner of each page bears a printed blood-red splotch, symbolising martyrdom. There are photographs of the Ayatollah in the midst of adoring masses. On the back cover is emblazoned the message

> WE WANT TO LIBERATE JERUSALEM. WE CAN DO IT THROUGH THE LIBERATION OF IRAQ AND THE DESTRUCTION OF THE BAATH PARTY.

The theme of holy war pervades the text.

In Iraq the war against Iran is made out as the continuation of the Battle of Kadisiya in 637 when the Arabs destroyed the Sassanids' neo-Persian empire.

During 1984 the official Iraqi propaganda line was that Iraq was fighting for the Arab race against "the vile Persian aggressors". This language reflects an anti-Iranian feeling even among Iraqi Shi'as, who belong to the same sect as the enemy. The Hussein administration has taken great pains not to insult Shi'a Islam and therefore alienate its own Shi'as.

Iraqi broadcasts, at their mildest, favour the epithet "Magi"—referring to an unsavoury priestly caste in Persia. A message from Iraq's Third Army commander to President Hussein, and broadcast in Arabic and Persian (Farsi), stated:

> "Great sire! We gladly inform you of the annihilation of thousands of harmful Magi insects that carried out an abortive offensive last night. . . . We have turned the Magi into black masses piled upon our pure soil. . . . May Allah grant us success!"

In the last year the Hussein regime has been making increasing use of Koranic verse and metaphor in addressing its people and to justify the war on religious grounds. In references to Iran, the stress is on Teheran's squandering of its own youth in doomed suicide assaults. This theme is driven home in Baghdad's Persian-language broadcasts.

During 1984–86 Iraqi propagandists tried to impress journalists with displays of masses of Iranian corpses. Photographers were encouraged to photograph them from every angle. The campaign had three aims: to reassure public opinion at home which was increasingly disturbed by an unending war; to proclaim the inevitability of an Iraqi victory to the world outside; to demoralise the enemy Iranians by highlighting their casualties.

Every evening Baghdad television beamed through powerful relay transmitters a programme in Persian, with Arabic sub-titles to Iranians living near the border. The programme consisted of 60 minutes of camera coverage of corpses. The commentary was invariably the same—"Khomeini and his akhunds (clergy) are sending your sons to be butchered pointlessly".

"Dawn Eight" and "Dawn Nine"

Iran began a major amphibious assault, code-named Dawn Eight, on 9 February 1986. Predictably, it came on the eve of the seventh anniversary of the Islamic revolution in Iran. Troops crossed the Shatt al-Arab at several points south-east of Basra and captured the coastal town of Faw. Iraqi resistance stiffened and the Seventh and Third Corps contained the five divisions committed by the Iranians. Casualties were heavy. The Iraqis claimed to have killed 25,000 Iranians while the Iranians claimed 12,000 Iraqi dead. The élite Iraqi Republican Guard Division was wiped out. The Iranian offensive approached so close to the Kuwaiti border that Kuwait and Saudi Arabia put their forces on high alert.

On 24–25 February 1986 the Iranians launched their Dawn Nine offensive, 500 miles north of the Dawn Eight Front. This second attack, in line with Iranian strategy of forcing the Iraqis to guess where the main assault might fall, was north-east of Sulaimaniya, about 60 miles east of the vital Kirkuk oil production centre. The Iranians captured about 50 deserted Kurdish villages and 200 square miles of territory. Dawn Nine was largely designed to encourage the pro-Iranian Kurdish guerrillas, who made many attacks against Iraqi targets, even army bases.

Yet again, Iraq used mustard gas, the nerve gas tabun and cyanide-based poison. The Iranian high command claimed that 8,500 of its soldiers had been affected. The Iraqis denied using chemical weapons but a United Nations specialist medical team of doctors from Spain, Australia, Sweden and Switzerland reported that gas had killed or seriously wounded at least 700 Iranian soldiers. The Iranians, who were more shocked by the gas attacks than by the death in battle of tens of thousands of their boy martyrs, were encouraged by the report. Foreign diplomats in Teheran regarded it as a moral victory for Iran and said that Iran was now less likely to use chemical weapons in retaliation. The report was expected to have an adverse effect on Iraq's relations with the West, particularly the United States and France.

At the end of Dawn Eight and Dawn Nine the Iranians could claim a victory in the north. On the central front Iranian troops in the Meimak Hills were less than 70 miles from Baghdad. In the Hawizah Marshes, Iranian soldiers were within sight of the Baghdad-Basra highway. On the Faw peninsula they clung to their positions. But with Iraqi aircraft dominating all the battlefronts a significant Iranian breakthrough still seemed unlikely.

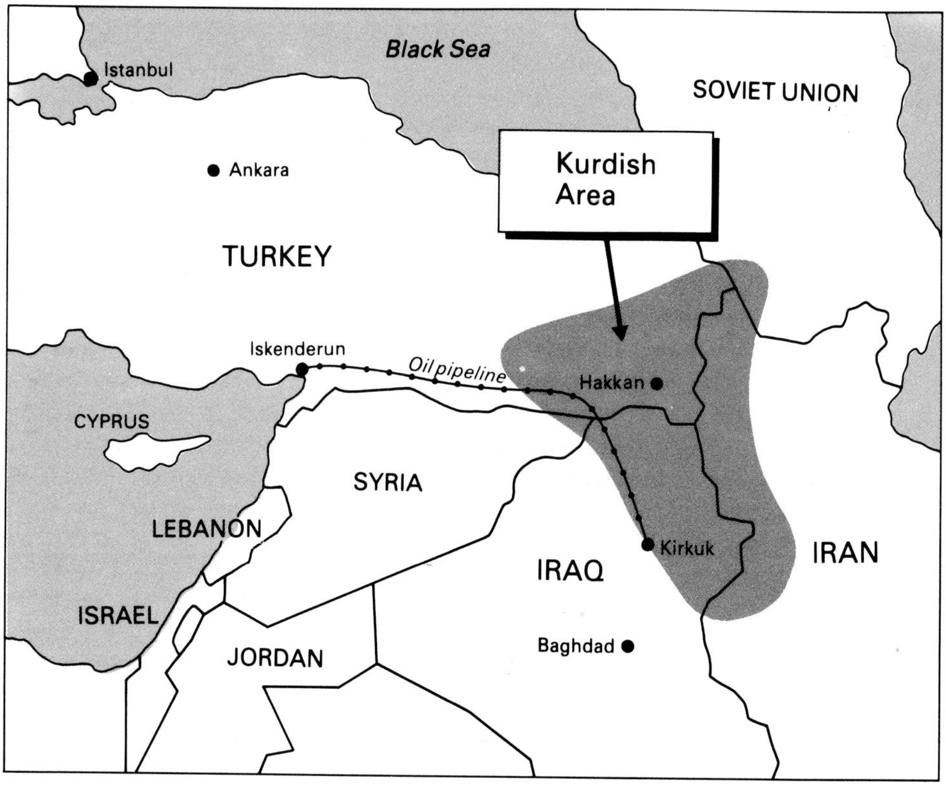

Kurdish Guerrilla War

Kurdish Guerrilla War

WITHOUT ALLIES

Background

The Kurds of Iraq, Iran, Turkey and Syria have long wanted a homeland, and since the early 1970s have fought for it. There are two major reasons for their failure—they cannot match the military strength of the national armies of the four countries in which they operate, and they have no unified command. Each political group, especially the Iranian and Iraqi Kurds, has been intent on benefits and gains for itself alone. In the early 1970s Mullah Mustafa Barzani, leader of the Iraqi Kurdistan Democratic Party (KDP), allied himself to the Shah of Iran and fought against the Iranian Kurds. In 1975 the Shah withdrew his support for the KDP in return for territorial concessions by Baghdad; the KDP Kurds then found themselves caught between the Iranian and Iraqi armies.

When the KDP revolt collapsed in 1975, the Iraqi army destroyed all Kurdish villages within 20 km of the Turkish border and deported their populations to southern Iraq. Later they were allowed to return—to refuge camps where they could be "controlled". Many have now left the camps.

In 1983, with the Iran-Iraq Gulf War at its height, the KDP helped the Iranians cross into Iraq. At that time the rival Patriotic Union of Kurdistan (PUK) was fighting Iraqi troops in Iraqi Kurdistan and helping to train the Iranian Kurds to fight Khomeini's troops in Iran.

The War in 1985

Late in 1984 the KDP, under Iranian control, started guerrilla activities against Iraq in the border areas of Iraqi Kurdistan and attracted new recruits. The KDP's leader is Massoud Barzani, son of Mustafa, the former chief. Barzani's decision is based on his belief that Iran will win the Gulf War and that the KDP had better be on the winning side.

The KDP has about 15,000 men in commands of 1,500 each; they co-operate with the Iraqi Army, but do not take part in operations with them. Generally, they prefer hit-and-run guerrilla operations. Unlike their Iraqi army allies, they are not fighting to capture ground. They hit a police station, a "security intelligence" post or a supply dump and are out of the way before the Iraqis can react.

The organisation of the KDP is interesting. The guerrillas operate in sections or *dastas* of 12, and when a dasta has been given an assignment its leader will walk the men from the mountain base to a village perhaps a week's travel away. In this village they meet secret KDP members who provide all the logistical help needed and from here they plan the details of their operation, which could be in an Iraqi city. Openly moving on to the city, they obtain weapons, generally Israeli Uzi submachine guns, from other agents. The support system is so efficient that a *dasta* can stay in a city for up to 10 days.

Several members of the Iraqi Intelligence service have been murdered by Kurds without any arrests being made. Killing Intelligence agents is calculated to show the Iraqi regime that the Kurds' own Intelligence is good, and that they can penetrate areas of tight security. Some Kurds have infiltrated the Iraqi Army, but their cover is so good that few are caught.

The Kurds based under cover in Kirkuk are operating against the great oil installations there. The pipelines are heavily guarded with checkpoints and barbed wire preventing anybody from getting within 200 m. Even so, the Kurds do blow holes in the pipeline. They have attempted to damage the main oil plants, but as all authorised people are kept at least 10 km away, this is difficult.

Occasionally the KDP mounts a conventional assault. On one occasion in June 1985, in a night attack, a 50-strong patrol mortared a 400-strong army unit protecting the Said Sikkik–Khornawazan road and took prisoners. The soldiers suffered heavy casualties and the survivors were routed.

By the end of 1985 the KDP controlled an area 200 km long by 25 km wide along the Turkish border and stretching from Syria to near Barzan. This is not a vast area, but it is large enough for camouflaged bases, the radio station and headquarters to be safe. At least 11,000 guerrillas are in the region, 6,000 from the KDP and smaller numbers from the Iraqi Communist Party, the Socialist Party of Kurdistan, the Iraqi Dawe Party (a Shi'a group) and the Turkish Workers' Party of Kurdistan. All are members of the National Democratic Front.

This area under guerrilla control is strategically important. Through it passes the international highway linking Iraq with Turkey and Europe and the main oil pipeline. The defences are formidable and include minefields, dog patrols and strongpoints so well made that an air attack would be needed to destroy them; the guerrillas have no aircraft. There is another deterrent—the KDP's fear of Turkish Army reprisals. At least 40,000 Turks are stationed along the border, and they would certainly destroy the guerrillas if they blow up the oil pipeline.

The KDP's daring exploits in the cities and the success of its *dastas* in many parts of Iraq are important, but more important still is that the guerrillas tie down thousands of Iraqi troops in Kurdistan when the Saddam Hussein regime needs every man it can get to fight Iran a long way to the south.

Some guerrillas in the Kurds "no-go" area are deserters from the Iraqi Army; these men invariably bring their weapons with them. One Kurdish deserter drove a new APC into the guerrilla camp.

The KDP is not short of arms and ammunition; supplies come not only from Iran but from Syria and Libya, who are Iran's allies in its war against Iraq. Main personal weapons are Kalashnikovs and rocket-propelled grenade launchers. They have large stocks of mines and explosives.

Few foreigners are helping the guerrillas, but small numbers of British, Americans, French and Italians are known to be operating as "advisers". The KDP movement is recruiting boys as young as 14, though only in an emergency are they allowed to fight before they are 18. The instructors say that 4 years are needed to make a guerrilla "who can stay alive".

Barzani and the other guerrilla leaders know that the "Iranian connection" under Khomeini could collapse in the same way as that under the Shah. They back the connection because no other is on offer to them. Their great hope is that the Iranians, having defeated Iraq, will go to war against Turkey in an effort to bring it back into fundamentalist Islam. Then, say the KDP leaders, there is a chance that Iran will give them a homeland.

Iraq's "Secret" Report

In 1986 a military/political report ordered by Saddam Hussein, which was supposed to be kept secret, showed that the KDP had tripled that part of Iraq under its control. The guerrillas had drawn 160,000 Iraqi soldiers away from the Iranian battlefront. The report stated that the morale of the guerrillas was high and that their supply of weapons had "improved". As the supply was already more than adequate this assessment was alarming.

The officers responsible for the report stated that the Kurds would "soon" be able to make frequent raids on the road from Baghdad to Turkey and the oil pipeline that runs parallel to it. "If the Kurds are allowed to continue to expand their operations they will hit our targets near Kirkuk," the report warned. Kirkuk is the Iraqi oil centre, though its main installations will probably remain safe from attack. To attack and hopefully to capture Kirkuk is one of the Kurds' main ambitions; they insist the city is theirs. The reporting officers found some consolation in telling Saddam Hussein that the Kurds were as divided as ever and advised political and propaganda operations to divide them still further.

No matter what Iraq does it may find that it has to pay a higher price than any other nation for the unfulfilled promise of Kurdish self-determination that was made by the great powers after World War 1.

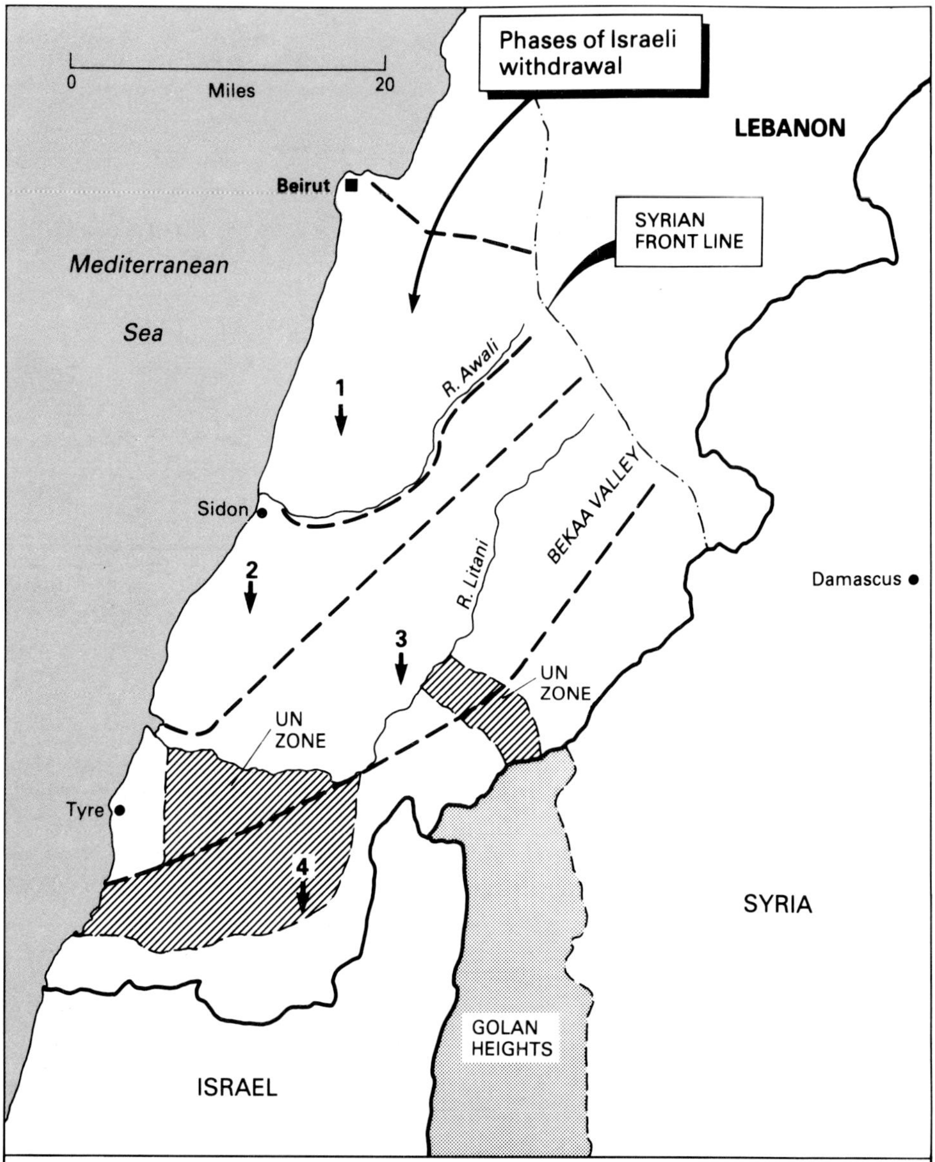

1: Israeli troops pull out of Beirut area in September 1983 and establish new front line at the Awali. **2:** Israel withdraws from Awali in February 1985. **3:** Third pullback in April. **4:** Some of this area remains a security zone for Israel.

War in Lebanon

War in Lebanon

RELIGION AND RIVALRY

Background

The conflict in Lebanon began in 1975 as a civil war, mainly between the "Lebanese Forces", the name adopted by the Maronite Christian militia, and the terrorists of the PLO. The PLO, having settled in southern Beirut after their expulsion from Jordan in 1970, refused to obey Lebanese laws and regulations and became a state within a state. Their region became known as Fatahland, after the name of the PLO faction commanded personally by Yasser Arafat, chairman of the PLO.

When the Christians appeared to be losing to the PLO, the Syrian Army went to their help, but later the Syrians transferred their allegiance to the PLO.

Lebanese politics had always been finely balanced between Christians and Muslims, but the presence of the predominantly Muslim PLO gave the country a Muslim majority. The PLO bullied the Shi'a Muslims of the south and used the region as a base from which to launch terrorists attacks across the border against Israel. When the Israelis retaliated, the Shi'a people were sometimes caught in the crossfire.

The general tension affected the Sunni Muslims of central-west Lebanon and the Druse of the hills, notably in the Shouf Mountains, south of Beirut. The Syrian government moved in a "peace-keeping force" which was really an army of occupation; its units were scattered throughout Lebanon, but were predominantly in the Bekaa Valley, eastern Lebanon, from where they attacked Christian centres. Syria had never recognised Lebanon as an independent country and wanted to absorb it (with Jordan) into a "Greater Syria".

Fighting continued intermittently for the next 8 years and at least 100,000 people were killed, many of them in massacres. In 1978 an Israeli column invaded southern Lebanon against the PLO in Operation Litani, in retaliation against PLO raids and harassment. The UN sent in a multinational force, the United Nations Interim Force in Lebanon (UNIFIL), which was supposed to prevent PLO incursions into Israel. It failed to do this.

On 6 June 1982 Israel launched Operation Peace for Galilee. It was designed to destroy the PLO bases and clear the terrorists from a 40-km zone next to the Israeli border. Since Syria was aiding the PLO—one PLO faction, Saiqa, was Syrian-led and financed—and because Syrian missiles positioned in the Bekaa were seen as a dangerous threat to Israel itself, the Israel Defence Forces (IDF) clashed with the Syrian Army and Air Force.

The Shi'a people of southern Lebanon welcomed their Israeli liberators and Israel decisively won the war. By October 1982 the PLO organisation was destroyed and the survivors evacuated Beirut, the PLO "capital". The Syrians lost more than 100 aircraft and many tanks. The Israelis' material losses were light, but they lost 354 men killed. The IDF had liberated southern Lebanon, but Israeli leaders felt compelled to keep an army of occupation in the area to counter PLO attempts to move back and Syrian expansion and to keep the peace between the warring Christian, Muslims and Druse. Even so, the IDF failed to prevent the massacre, in Sabra and Shatila refugee camps, of at least 800 Palestinians by Christian militiamen.

In May 1983 Israel and Lebanon signed a peace agreement, but Syrian pressure on Lebanon destroyed this accord. By now, too, Iranian fundamentalists had established themselves in Baalbek from where they incited the Shi'a Lebanese Shi'as to violence in the cause of holy war. At this time a bitter conflict had developed between the pro- and anti-Arafat factions of the PLO; the Fatah commander in the Bekaa, Musa Awad (Abu Musa), proclaimed a mutiny against Arafat's leadership. Supported by about half the PLO fighters, he was recognised and backed by Syria.

In an attempt to bring stability to Beirut at least, a Multinational Force was organised, comprising US marines, French paratroops, Italian soldiers and a British contingent of about 100.

On 23 October 1983 a suicide attacker rammed a truck laden with 2000 lb of explosives into the building which housed the headquarters of the US Marine Corps in Beirut; 241 marines were killed in the explosion. Almost simultaneously, another suicide truck bomber crashed into a French base and 58 paratroops died. On 4 November a similar attack was made on the Israeli Border Police HQ in Tyre and 60 people, most of them Muslims, were crushed to death. The MNF pulled out of Beirut. Some of the best-trained troops in the world, with some of the best equipment and sophisticated logistical support, had been forced out of their positions—and out of Lebanon altogether—by a handful of primitives whose principal weapons were fanaticism and a desire for martyrdom.

The MNF had failed because neither its political nor military leaders had comprehended the nature of "peace-keeping" or of warfare in this area. The MNF's conventional approach had been useless against fanaticism.

Meanwhile, another part of the long-running war was going on in northern Lebanon, where Fatah troops loyal to Yasser Arafat were fighting those now under Syrian control. Besieged in Tripoli, Arafat and his supporters came under heavy fire from PLO mutineers backed by Syrian armour and artillery. The internecine fighting was some of the most savage seen in Lebanon, and many Lebanese civilians were killed. Finally, about 4,000 pro-Arafat men were evacuated from Tripoli by sea.

In the south, Israel, unlike the US, could not afford the luxury of what the Americans called "redeployment". A military withdrawal from Lebanon would be interpreted by the Arabs as retreat or defeat. At the very least, Israel's deterrent stance would be seriously eroded. The Arabs would focus on war, not diplomacy, if they perceived a "weak" and defeated Israel.

Some Israeli war analysts claimed that getting out of Lebanon would not weaken the nation's deterrent posture but restore it; the IDF could then devote its energies to the kind of training and reconstruction of its fighting forces which would ensure Israel's deterrent capabilities.

But the MNF force's withdrawal and the disappearance of the massive American battleships from the coast of Lebanon left Israel with a massive handicap. Western

Israeli infantry in action against Shi'a fighters in southern Lebanon during the summer of 1985.

prestige collapsed. The Syrians, Lebanese, PLO and much of the Arab world now knew that the Western powers were impotent in the Middle East. This quickly led to the conviction that Israel had no supporters and was therefore more vulnerable to attack. This was particularly clear to the Iranian *agents provacateur* in Baalbek and to the Syrian leaders who incited the Shi'a Muslims of the south to attack the IDF. As early as March 1984 70% of anti-Israeli activity in the south was Shi'a, not Palestinian.

Israeli efforts to prevent the infiltration of terrorists and sabotage material involved stringent inspection arrangements at the Awali River bridges. This disrupted the flow of traffic between Beirut and the south and greatly inconvenienced the inhabitants of southern Lebanon. Anti-Israeli sentiment was further inflamed by normal searches and road blocks.

The strategy of the Syrian–Iranian alliance was to force the Israelis to pull back without any corresponding Syrian withdrawals and without any risk to themselves. Their tactics were to use the Shi'as and *their* tactics were to harass the Israeli troops with standard guerrilla hit-and-run attacks, with booby traps and with suicide attacks. Israeli casualties in Tyre, Sidon, Nabatiye and elsewhere mounted.

The Conflict from 1985

In September 1984 Shimon Peres took office as Prime Minister of Israel following an election which produced no clear victory either for his Labour Party or for the Likud coalition. He had campaigned on a promise to withdraw Israeli soldiers from Lebanon as soon as possible, a policy which reflected public opinion.

The phased disengagement and withdrawal, a classic of its type, was one of the most difficult in military history. Apart from the need to find a practical future security policy, Israel had a responsibility to keep the peace in areas under its control. Military Intelligence warned that when the IDF moved out a dangerous vacuum would be created; Palestinian refugees would become the victims of Shi'a murder squads and Christian villages in southern Lebanon would also be vulnerable. Sidon was a particular problem, with 80,000 Shi'as, 45,000 Sunnis, 20,000 Christians and 80,000 Palestinians living in and around the city. Each of these groups was swearing vengeance against all the others, while the Druse, in the hills, were also waiting for the opportunity to attack Christians in particular.

The Lebanese Army was weak and split by sectarian hatreds—some units were Muslim and others Christian. The Lebanese government would not order these units into areas to be evacuated by the Israelis. Syria vetoed any UN military presence in the area because it might have prevented conflict; the Syrians needed conflict in line with their policy of dividing and ruling.

Despite the deliberately steady pace of the Israeli withdrawal, Syrian, Iranian and Lebanese Shi'a propaganda claimed that the Israelis were fleeing. As was only to be expected, the propagandists told the Shi'a villages that the Americans and French had already been driven out of Lebanon—now it was the time to kill Israelis. At times during the pullback, attacks on Israeli troops reached 10 or 12 a day. A Shi'a suicide bomber drove a pick-up truck loaded with 220 lb of explosives into a lorry loaded with Israeli soldiers going home; 12 of them died in the explosion.

In April 1985 the Syrian Popular Party in Lebanon recruited a 16-year-old girl, Sana Mhaydali, as a suicide bomber. She could not drive a car, but her recruiter gave her an elementary lesson in putting a car into gear and steering it, then they push-started it for Sana's last journey. She rammed an Israeli military vehicle, killing two Israeli soldiers and herself.

Another 16-year-old Lebanese Shi'a, Muhammad Burro, a trained suicide car bomber, was captured during an Israeli raid on a southern village. He told the *New York Times* that he had been blackmailed into volunteering for martyrdom. Abu Hassan, security chief for Amal in Beirut, had threatened to make a lot of trouble for the Burro family who were already deep in financial and other difficulties.

In mid-April 1985 Christian artillery east of Sidon shelled Lebanese Muslim and Palestinian areas of the city; Muslim gunners responded and fighting went on for 2 weeks. The Christian gunners withdrew, leaving the Christian villagers at the mercy of the Muslims who drove them from their homes, thus causing another exodus of refugees.

Even bloodier fighting occurred in West Beirut, which on the night of 16 April became a battlefield yet again. The Sunni Muslims had only one militia, the Murabitoun. For a year PLO fighters loyal to Yasser Arafat had been trickling back into West Beirut. Simultaneously, Arafat had been sending money and arms to the Murabitoun with which he planned an alliance. The Syrian leadership and their Shi'a Amal allies had to crush this union before it became an effective political and military force.

After several hours' fighting between Sunnis and Shi'as, the battle to control the square of Sakiet el Janzir was still undecided. At this point Syrian Intelligence officers sent in the Druse militia on the side of Amal; by mid-morning the Murabitoun was destroyed. Hundreds of homes were looted and several families were found massacred in their homes.

Despite the guerrilla attacks, the Israeli withdrawal, under Major-General Or Orr and Brigadier Chaim Erez, was effected without changes of plan. Computers determined the number of trucks and man-hours needed for the massive operation. The army planners faced enormous problems in getting heavy equipment and static installations dismantled and returned to Israel. Whatever could not be moved was destroyed and ploughed into the ground. Deep tunnels dug by the PLO in the mountains east of Sidon, and other bunker and trench complexes, were demolished to make it more difficult for the PLO to return to its former territory.

The most difficult aspect of the evacuation was providing security for the long heavy convoys moving through guerrilla-infested countryside. Numbers of jeeps armed with two heavy machine-guns guarded each convoy, while foot patrols in flak jackets searched the sides of the roads for bombs and land mines. All intersections were sealed off with concertina barbed wire and armed half-tracks as convoys passed. As they left, the Israelis distributed leaflets which warned the Shi'as against terrorist acts and threatened retaliation; this was part of the "Iron Fist" policy evolved by the Israeli Defence Minister, Yitzhak Rabin.

The role of UNIFIL was hesitant and ambiguous during the Israeli withdrawal, and its own officers were suggesting that it should be pulled out. For 7 years it had failed to carry out the duties laid down for it by the Security Council, and between 1982 and 1985 UNIFIL soldiers could do little more than provide some humanitarian assistance and limited protection to the civilian population.

UNIFIL troops are under orders not to shoot except in direct self-defence. This has never made sense to the Lebanese of southern Lebanon, where every man and boy is armed and ready to shoot on the slightest provocation or suspicion. Trying to keep the peace was dangerous and in the period 1978-1986 44 UN troops were killed in action.

The IDF left UNIFIL out of account when preparing its defences along the Lebanon border. The security plan was based on a buffer zone between a mile and a half and 10 miles in depth, depending on the terrain, which was to be patrolled by the South Lebanon Army under Major-General Antoine Lahad. On the Lebanese side of the border IDF engineers dug an anti-vehicle ditch designed to swallow up any suicide attacker and his car-bomb. The ditch is not continuous, but in places runs for miles. For a length of 55 miles watchtowers, electronic warning devices and floodlights are among the defensive devices. Well-trained regular soldiers rather than reservists man the defences.

In May and June 1985 a fierce, almost continuous, battle took place between Shi'a Amal fighters and PLO returnees for possession of the refugeee camps Sabra, Shatila and Bourg el-Baranje in south Beirut. These camps are really suburbs with many alleys, cellars and tunnels, where the Palestinians held out desperately. The Shi'as massacred many civilians, including women and children and hospital staff. It is very likely that as many people died in these massacres as in the 1982 Sabra–Shatila slaughter.

After having removed vast amounts of equipment, the Israelis left the southern Bekaa Valley where they and the Syrian Army had confronted each other for 3 years. With sophisticated surveillance equipment, including mini Remotely Piloted Vehicles (RPV), the IDF believes that it could detect any Syrian offensive movement and that possession of high ground was no longer necessary. The RPV is the so-called "spy in the sky", a small pilotless plane with a television camera which sends back to its command post "real time" intelligence of an enemy's activities.

During its 3 years in Lebanon, the IDF had lost 654 men killed, but it had destroyed the PLO and it had demonstrated to Syria, Israel's hostile neighbour, that an invasion of Israel would be defeated. Nevertheless, it was widely criticised in Israel; many people questioned the necessity for the war and the long occupation of Lebanon. The IDF, it was said, excelled in quick-retaliation campaigns and should never be used in siege warfare, as in Beirut, or in occupation, as in southern Lebanon.

For the military analyst, professional or amateur, Israeli's 3-year war in Lebanon is interesting for several reasons.

—The air force with the best reconnaissance wins. With supremely efficient technology, including the RPV, the IDF was able to knock out the Syrian anti-aircraft defences and to inflict overwhelming losses on the Syrian Air Force.

—Israel's Merkava tank, first used in Lebanon in 1982, is the best in the Middle East. Though slower than some other tanks, the Merkava has a high-standard main gun and excellent ammunition. Its rear entry and exit hatch provide the crew with considerably more protection than with a top-hatch tank.

—A terrorist-guerrilla force which seeks to make itself into a quasi-regular army can be quickly driven from its area of operations, even when it is ideally suited for guerrilla activity. This was one of the PLO's most serious mistakes.

—Military occupation, even when protective and "benign", always arouses resistance. The Israelis had not learned this lesson despite the experiences of the

Americans in Vietnam, the Russians in Afghanistan and the Cubans in Angola, and they stayed too long in Lebanon.

—It became clear from Israel vs Syria air operations that in any future Middle East war attack helicopters with advanced weapon systems and far greater operational latitude will be more widely used.

At the end of 1985 the Syrians had replaced all their 1982 losses and were actually in a better position, since they had received newer planes for the older ones which they had lost. They had also increased the number of surface-to-air missiles of all varieties. Syria's deployment of SA-5 missiles covered large areas of Israel's air space and had the potential for limiting Israeli Air Force operations.

However, by 1986 the Israeli Air Force had gained a qualitative advantage with the production of their own Lavi combat aircraft. The plane's class and mission role is similar to that of the American F-16, but designed for Israeli Air Force requirements.

During the latter part of 1985 serious and bloody conflicts took place in and around the northern Lebanese city of Tripoli. Sunni Muslim fundamentalists of the pro-Palestinian Taweed organisation or Islamic Unification Movement came under attack from a variety of Islamic militias, backed by the Syrians. Fighting was so severe that in September–October 1985 two-thirds of the population of 500,000 moved out as refugees; at least 500 people were killed and 1,000 injured. This phase of the long-running Lebanon war was probably more intense and bitter than at any time during the long war. An Islamic group kidnapped Soviet Beirut Embassy officials in an attempt to force the Russians to put pressure on Syria to stop attacking the Sunnis; some of these hostages were killed when the Syrians did not restrict the fighting. There was no doubt by this time that Syria dominated every aspect of Lebanese political life and could create, moderate or stop any of the fighting.

Preparing for Another War

During the early part of 1986 Syria seemed to be preparing for a war against Israel on the Golan Heights. President Assad twice said, in public addresses, that "the Golan will be located in the heart of Syria and not on its border". This was a thinly veiled threat of invasion of Israel. More directly, Assad announced, "We have a date, soon, with our brothers in the Golan, in Palestine and in southern Lebanon."

At that time—March-April 1986—Syria was close to completing a massive military build-up along the Golan border with Israel. Six Syrian divisions, each about 60,000 strong, were stationed in the area and many others could be quickly brought in. Nine batteries of SA-6 Soviet anti-aircraft batteries were in position close to the demarcation line between Syria and the Golan; this was a breach of the 1974 agreement signed by Syria and Israel following the 1973 war.

An air-mobile force of several battalions had been added to the enormous army of 470,000; rather like commandos, they are trained to operate from helicopter gunships. The outmoded Soviet T-54 tanks had been replaced with the advanced T-72 and even the T-72 Super. About 2,000 of these tanks—out of 4,000 tanks in all—were deployed between Damascus and the Golan Heights.

Syria's armed forces are kept permanently on high-alert status; thus, Syria could mount a major offensive within a few hours.

VENDA
"Independent" homelands
LEBOWA
Other homelands
ZIMBABWE
VENDA
MOZAMBIQUE
GAZANKULU
BOTSWANA
LEBOWA
TRANSVAAL
SOUTH NDEBELE
Pretoria
Johannesburg
KANGWANE
SWAZILAND
ORANGE FREE STATE
QWAQWA
BOPHUTHATSWANA
NATAL
KWAZULU
Maseru
LESOTHO
Durban
SOUTH AFRICA
TRANSKEI
CAPE PROVINCE
INDIAN OCEAN
CISKEI

Lesotho

Coup in Lesotho

VICTORY FOR SOUTH AFRICA

From 1 January 1986 South Africa blockaded road and rail traffic into Lesotho, a constitutional monarchy reigned over by King Moshoeshoe II and ruled by Chief Jonathan, the prime minister.

A mountainous country of only 30,355 m^2 and with a population of 1.4 million, Lesotho had three political parties, Jonathan's ruling Basotho National Party, the Basutoland Congress Party, led by the self-exiled Ntsu Mokhahle, and the neutral royalist Maremthlou-Freedom Party. Most ordinary people resented Chief Jonathan's dictatorial and incompetent regime.

The South African blockade, which reduced food, fuel and other supplies, intensified tensions within Lesotho. It was clearly meant to have this effect; South African border guards told Lesotho citizens seeking to cross the border for work, to buy food or visit a doctor, to blame Chief Jonathan for their disappointment.

South Africa alleged that Lesotho under Chief Jonathan operated as a staging post for members of the African National Congress (ANC) which is outlawed by South Africa. This was true; ANC members flew in and out on the direct flight between Maseru in Lesotho and Maputo in Mozambique. The South Africans said that Lesotho, as a sovereign state, was free to choose its own diplomatic ties, but they did not approve of the presence in Maseru of the large staffs in the Soviet Union and North Korean embassies.

On 17 January General Justin Lekhanya, commander of Lesotho's 12,000-strong paramilitary force, met senior South African security officials in Pretoria. Lekhanya's coup took place on 20 January when his troops seized the barracks, radio station, airfield and railway stations. During the brief fighting between his men and those loyal to Chief Jonathan, 17 soldiers were killed by small-arms fire.

Immediately after General Lekhanya made his victory broadcast, South Africa lifted its blockade and its deputy-director general of foreign affairs, Neil Van Heerden, arrived in Maseru to arrange for "co-operation" between the two countries.

The major significance of the coup is its effect on the ANC, which suffered a serious setback in its plans for operations within South Africa. For instance, it was expecting aid from Cuba, channelled through Lesotho. Immediately before the coup a Cuban delegation under the Cuban Foreign Minister himself, Isidoro Malmierca, arrived in Lesotho. This aroused anxiety and anger in South Africa. General Lekhanya's ruling military council announced that the presence in Lesotho of ANC people was "totally unacceptable". South African officials had used the same phrase.

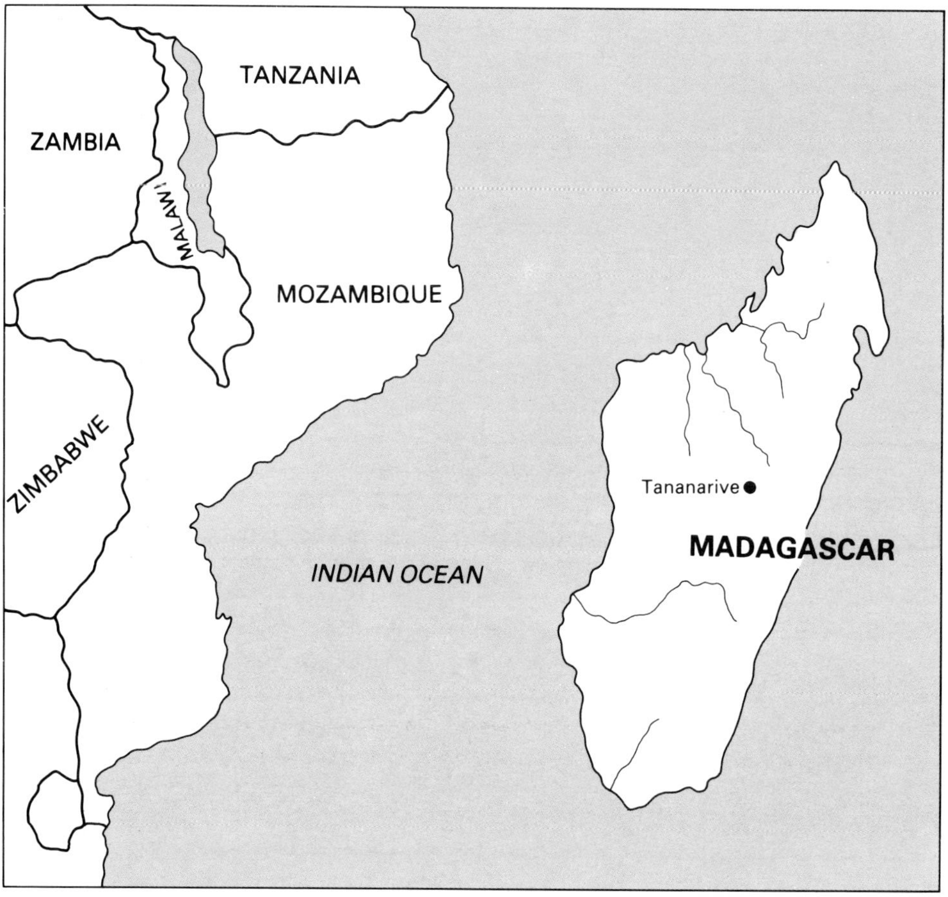

Madagascar Civil War

Madagascar Civil War

THE KUNG FU KILLERS

Just as it won election to the UN Security Council, the Indian Ocean island of Madagascar was faced with a political crisis by kung fu addicts opposed to the left-wing rule of President Didier Ratsiraka.

The many practitioners of kung fu among Madagascar's 9 million people went berserk after the police closed kung fu clubs following a government order banning the sport. The ban was imposed by Remi Tiandraza, Minister for Population, Social Conditions, Youth and Sport. Under his jurisdiction is the organisation Young Re-educated Malagasies, known locally as the TTS movement.

Since the mid-1970s its members, recruited from the island's poor, have been used as "punishment squads" and *agents provacateur* by President Ratsiraka's Arema party.

With the President's approval, Tiandraza took the step of banning kung fu because the craze for martial arts was sweeping Madagascar; the government feared that it could become dangerous. All the political parties, including the TTS, were training their fighting squads in kung fu. In fact, the kung fu clubs were replacing the secret societies which flourished in Madagascar during colonial times.

Tiandraza banned kung fu on 5 September 1984. After brooding for 3 months, angry mobs of kung fu fanatics attacked the headquarters of TTS in the capital, Tananarive. They slaughtered many victims with their bare hands, impaled heads on spikes around the railway station and dangled corpses from lamp standards. More than 100 people died in the slaughter, one of the worst in Madagascar's post-colonial turbulent history.

Ratsiraka, a former naval officer who had held power by sometimes brutal methods since 1975, dismissed the senior officers of the security forces for failing to keep order. The gendarmerie and the People's Armed Forces were given new commanders-in-chief.

The uprising might have been sparked off by the kung fu ban, but it reflects the tensions which are leading Madagascar towards civil war. Since 1975 Ratsiraka has nationalised banks and most industries and allied Madagascar to the Soviet Union, from where he obtains money and arms. He cut all the old ties with France.

Madagascar's main exports—coffee, cloves and vanilla—have suffered from erratic world markets, and Madagascar has enormous foreign debts.

Throughout the island during 1985 kung fu clubs met illegally and have been exploited by the opposition parties who see the kung fu people as the advance guard of a revolution. Police and soldiers have been found murdered by kung fu and other martial arts methods.

Kung fu was given great impetus in 1975 when some Bruce Lee films were shown in the capital. Now all films which include scenes relating to the martial arts are censored.

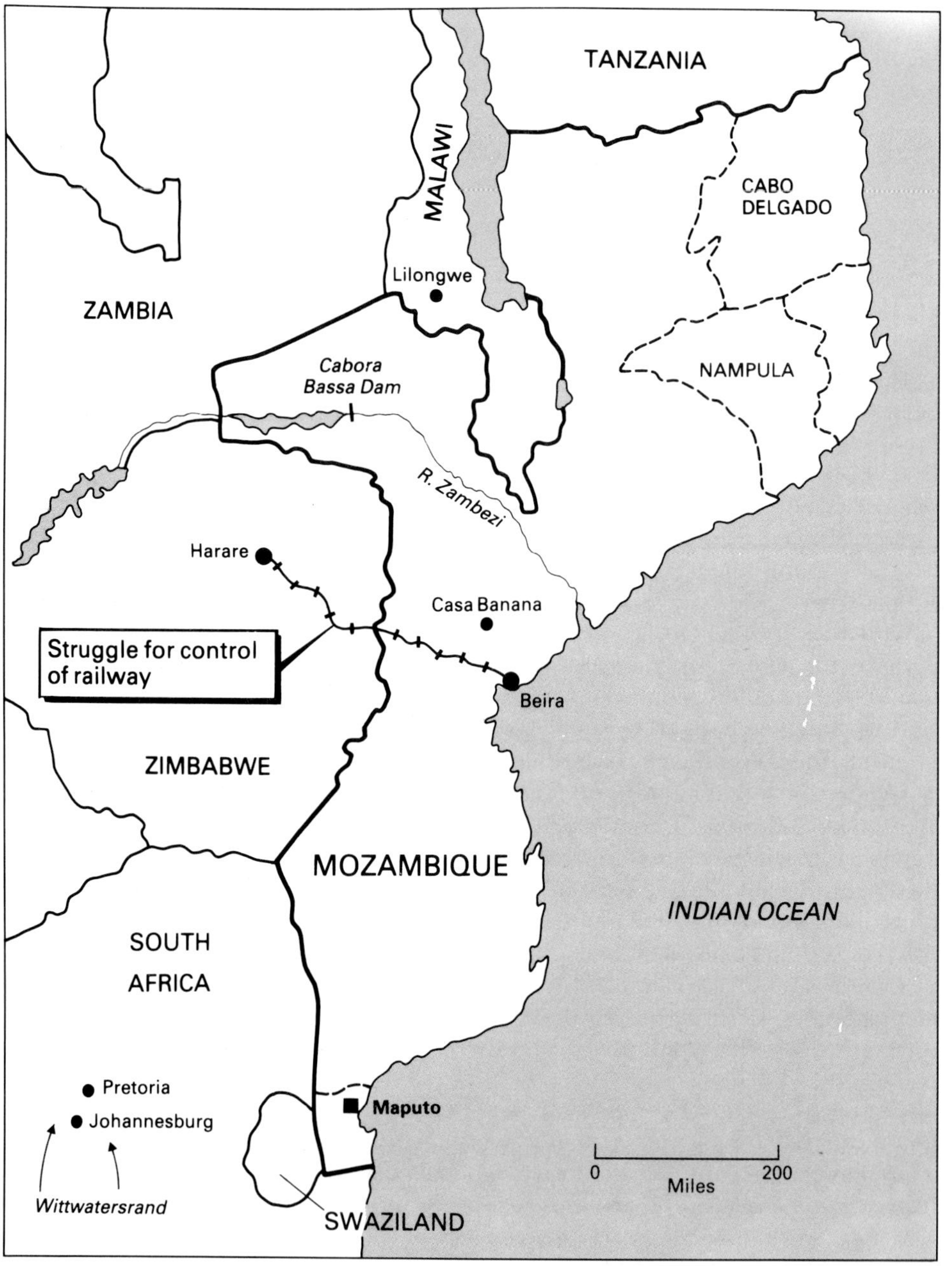

Mozambique Guerrilla War

Mozambique Guerrilla War

EQUALITY OF INCOMPETENCE

Background

In 1982 Mozambique, a former Portuguese colony, was a South African battlefield. Covert operations by South African military Intelligence were intent on ensuring that Mozambique could never become a threat to South African security. South African Defence Force (SADF) units raided the capital, Maputo; commandos blew up oil storage tanks at Beira, derailed trains and wrecked power lines.

South African Intelligence officers found a ready-made guerrilla resistance group to exploit. It was the MNR, better known as Renamo, which had been created in 1977 by the old pre-independence Rhodesian Central Intelligence Organisation. The white Rhodesians at that time were trying to reduce the amount of assistance which neighbouring black regimes might give to the Zimbabwean resistance.

The South Africans financed, trained, armed and transported the Renamo fighters. They justified this military activity by claiming that they were in "hot pursuit" of guerrillas of the African National Congress (ANC) who operated out of Maputo.

President Machel's incompetent Frelimo movement was powerless to combat Renamo. By 1982 about 90% of the country was in the hands of the Renamo guerrillas or rebel Frelimo commanders. Troops of all groups were little more than marauding bandits in a country hard hit by drought.

As a resistance movement, Renamo is ill led and incompetent and its members lack genuine resistance motivation. It is an amalgam of Portuguese expatriates and businessmen, dissident tribesmen, disgruntled Frelimo turncoats and mercenaries. Its nominal leader in Alfonso Dhlakama, while the Renamo spokesman is Eve Frenandes, who lives in Portugal.

By 1984 the South African leaders believed that they could better achieve their aim of keeping Mozambique out of the firing line against South Africa by making a peace pact with it. Accordingly, on 16 March, President Machel and the South African prime minister, P. W. Botha, signed the Nkomati Accord, a mutual non-aggression pact.

Machel hoped that South Africa would allow more Mozambicans into South Africa as workers—the number had dropped from 120,000 in the 1970s to 40,000 in 1982. In the old days, too, 500,000 South Africans a year flocked to Mozambique's Indian Ocean resorts, a tourist trade worth $14 million a year. The treaty, Machel hoped, would bring the South Africans back.

Nkomati was seen as a triumph for South African diplomacy; in fact it was simply a threat to Machel to get rid of the ANC. Before the end of 1984 President Machel was threatening to take South Africa to the international court for breaching the accord. Even so, as promised, Botha issued an order stopping all aid to Renamo—and made the chief of defence staff, General Constand Viljoen, personally responsible for seeing that the order was carried out. Machel kept part of his side of the bargain—the ANC closed down its activities in Mozambique.

The Americans became deeply involved in the period 1980-4 because they could see Mozambique going the way of Angola—into the Soviet–Cuban camp. With a 1,800-mile shoreline, an area of 600,000 square miles and a jagged interior border that runs from Tanzania in the north alongside Malawi, Zambia, Swaziland and South Africa, the territory is a key piece in the regional jigsaw.

The War in 1985

The guerrilla war continued, mainly because Renamo believed that it could eventually win. In any case, it was unreal to expect a ceasefire, because half of Renamo's units operate beyond command or control. SADF radio equipment originally supplied to Renamo was doctored to ensure that regional commanders could communicate only with the Transvaal, not with each other. In addition, before the Nkomati Accord SADF had flown in 2 years' supply of military supplies. The South Africans believed that if they withdrew their backing for Renamo it would wither away. They were mistaken. In fact, Renamo has a remarkable array of friends, including Brazilians, Portuguese, Moroccans, Saudis, Bavarians and Zaireans. During 1985 arms passed down the ancient Arab trade route from Oman, through Somalia and Zanzibar to the Mozambique coast. Much in the way of arms and equipment is flown to the Comores Islands and air-dropped to the Renamo fighters.

The Renamo leader in the Maputo area, Calisto Makuluve, took to calling his men "The President's Guard", a mocking title which underlines Renamo's threat to overthrow Machel.

The Soviet-trained armed forces of 15,500 have proved to be incapable of tackling the 9,000 guerrilla-bandits, who have terrorised large areas of the countryside. In addition, Mozambique is desperately short of helicopter gunships and transports, both essential for effective counter-insurgency operations.

However, in August 1985 Machel received military help from Zimbabwe. A force of 3,400 Zimbabwean soldiers had been doing guard duty on the 125-mile railway and oil pipeline which serve Zimbabwe from the Mozambique port of Beira. In secret, this force was built up to 13,500 to attack Renamo's Supreme Command base, known as Casa Banana, about 60 miles north-west of Beira.

It was a well-guarded base in heavy forest and virtually inaccessible by land. With its airstrip, communications and supplies, it was believed that the several thousands of rebels who used it could hold out for 3 years. The Zimbabwe Air Force bombed the place, paratroops landed on the airstrip and in a 5-hour strike they captured Casa Banana.

This was claimed, by Machel and his friend Robert Mugabe of Zimbabwe, as a great victory. In fact, Casa Banana was deserted when the Zimbabweans attacked and the Renamo officers left the communications centre a smouldering ruin. The attackers found large signs saying

WELCOME ZIMBABWE NATIONAL ARMY.

Renamo suffered fewer than 50 casualties. The Zimbabweans were shocked by their 159 casualties.

The victory was not decisive, because the Mozambique Army failed to back up the Zimbabwe troops, the government could not restore its own administration in the captured area, and the attackers did not discover the routes by which the great base was supplied. The capture of Casa Banana was a reverse, but the Renamo men quickly disengaged and disappeared into the bush.

Even so, Renamo owes more to government clumsiness and lack of security about its operations than to its own skill. Renamo is remarkable in that after 8 years as a guerrilla force it has no coherent leadership, no political philosophy and nothing to offer in place of the regime it seeks to destroy. Most of its members are brigands who are making food production even more difficult when large parts of Mozambique are suffering from a drought. Despite the dangers from the ill-disciplined Renamo, the giant British company Lonrho is planning large-scale agriculture in southern and central Mozambique.

Machel, a Marxist-Leninist, had promised to "create socialist men in Africa", but in 1985 he was turning his back on the calamitous system of state farms on the East European model in favour of private agriculture. Peasants are at last being given incentives to grow food. For the Russians, the unkindest cut of all has been Machel's decision to turn to Britain for the retraining of his middle rank officers in a programme for 1986-8; it will take place in Zimbabwe under the guidance of the British military training team there.

However, the East Germans are still a major force in Mozambique's intelligence agency, whose acronym is SNASP, and the Russians have their supporters in the circle of 15 people around the president who really run the country. Machel's position was seriously threatened in September 1985 when saboteurs blew up a munitions arsenal in Maputo, causing more than 100 casualties and destroying many buildings.

Machel's status *vis-à-vis* the superpowers and South Africa, the regional superpower, is difficult. He has a defence treaty with the Soviet Union and he still relies on Russian and Chinese military aid. Yet in 1985 he accepted an American offer of $1 million worth of supplies including army vehicles and uniforms. On top of all this, he has turned to South Africa for help because it is the most powerful country in the region and he would rather have it as a friend than an enemy. Neither the Russians, the Americans nor the South Africans trust him, and he has rivals for the leadership within his own party.

Caught between an incompetent army and an equally incompetent but brutal guerrilla/terrorist movement, the ordinary peasants suffer acutely. Probably 200,000 of a population of 12 million were murdered between 1977 and 1986.

Renamo claimed several successes in 1986, including the capture of Cavalo and the recapture of its airfield at Kangan'tol in February. Cavalo was the site of the headquarters of the combined Mozambique/Zimbabwe force which captured Renamo Gorongoso camp complex in August 1985. Renamo also overran Namacurra in Zambezia Province, killing 157 Mozambican and 59 Zimbabwean soldiers. In an attack on a bus travelling between the Swazi border and Maputo Renamo killed 7 people and wounded 20. One of the few government successes was the capture of the Mozambican National Resistance Movement's base at Manianje. Its greatest reverse was the loss of Casa Banana, which 400 guerrillas recaptured from 1,000 soldiers.

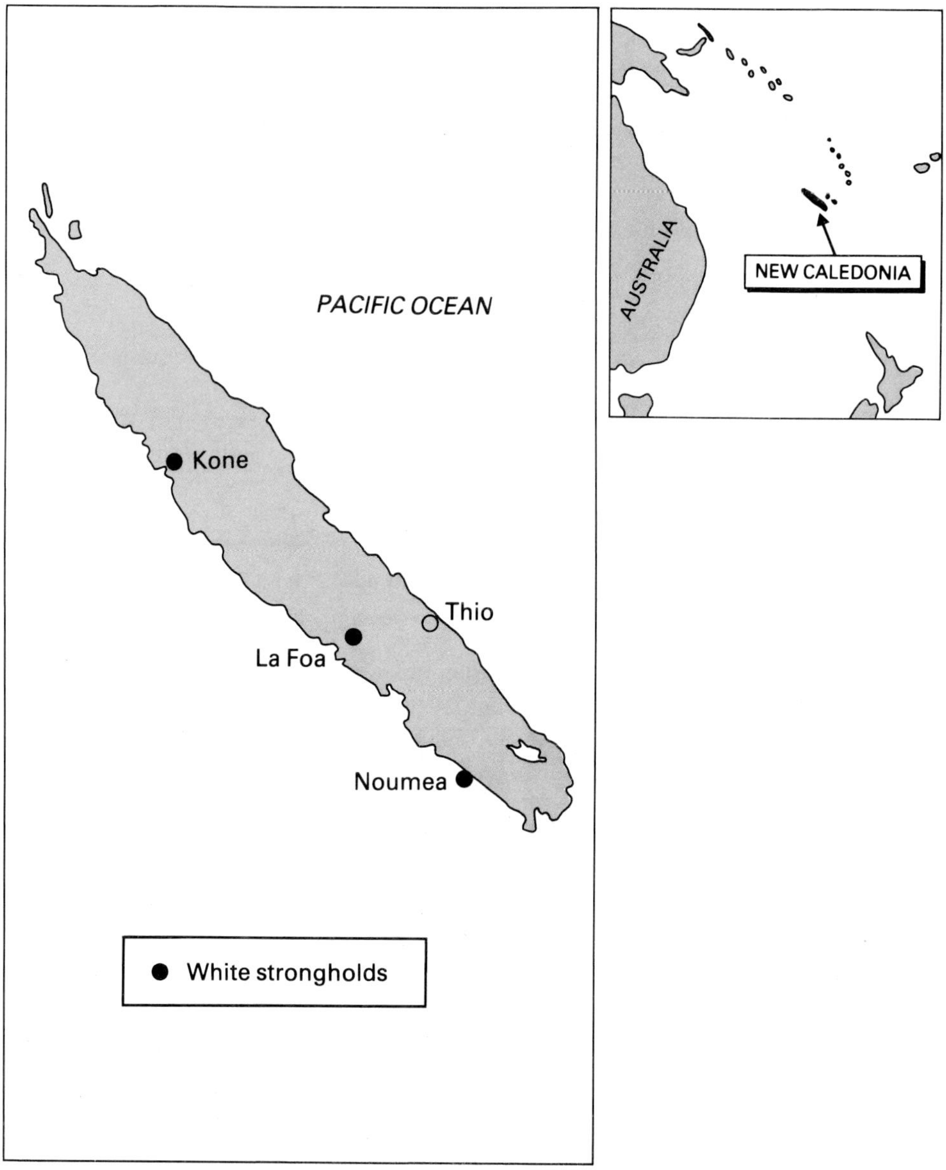

Civil War in New Caledonia

Civil War in New Caledonia

NO LONGER PARADISE

On 30 November 1984 a Melanesian black shot a French settler dead in his farmhouse and New Caledonia, a French possession long regarded as a Pacific paradise, gained an instant reputation as a trouble-spot.

In fact, the colony has known political unrest for many years. The indigenous Melanesians, the Kanaks, have long wanted independence. France, always reluctant to surrender its imperial possessions, wants the island's nickel deposits, on which the agreeable life of the capital, Noumea, is founded.

In November 1984 the Kanaks boycotted the local elections and formed a "provisional government". To try to calm the violence which followed, President Mitterrand flew to New Caledonia in January 1985 and appointed Edward Pisani as High Commissioner. Mr. Pisani brought in French troops and police but made it clear that his sympathies lay more with the Kanaks than the French settlers.

He induced the Kanaks to accept something less than full independence, but the two main white political parties blocked all his proposals; La Caledonia dans la Republique and the Front Caledonien want the status quo preserved.

French raiding parties killed about 30 Kanaks "to teach the native population a lesson". One of the victims was a Kanak leader, Elvi Machora, who had provoked the French by setting up road blocks which would not allow French-owned traffic to pass. His face now appears, Che Guevara style, on Kanak posters.

The Kanaks have reacted to French murder squads by forming death squads of their own. A French policeman was killed by a machete in the back, white houses have been burnt down and in the bush white cars are stoned.

The French flew in 6,700 peacekeeping soldiers and policemen, but they have had little to do. *Not* flying in are Australian and Japanese tourists, who were bringing wealth to the tourist industry.

The trouble in New Caledonia is reminiscent, in miniature, of the bloody days of the 1960s when Algeria was fighting for its independence; the Algerians' worst enemies were not the French Army, but the *colons*, the French settlers. Many of New Caledonia's *colons* came from Algeria and Indo-China and they talk of "fighting to the death this time".

The French Government proposes self-government, but with France retaining control of defence, the police and foreign policy. The newly arrived high commissioner, Fernand Wibaux—the "Minister for New Caledonia"—hoped that this formula would bring peace to the island.

Other countries have become involved in New Caledonia's affairs. Some Kanaks went to Libya in March 1985 for 2 weeks' arms training, and President Gaddafi is known to have offered the Kanak leaders arms and money. Some settlers allege that the neighbouring Melanesian countries of Papua-New Guinea and Vanuatu are sending arms to the New Caledonia guerrillas. Papua-New Guinea did send troops to help quell a rebellion, backed by whites, on the Vanuatu island of Espiritu Santo, but the prime minister denies involvement in New Caledonia.

The demography of New Caledonia is a problem for the French government. There are 62,000 Kanaks, 54,000 Europeans, mostly French, and 30,000 Asians and other Pacific islanders. In any referendum it is thought that the Asians and others will vote with the French to retain French sovereignty. The referendum will take place before the end of 1987.

Much depends on the 1986 parliamentary elections in France. A former Gaullist prime minister, M. Jacques Chaban-Delmas, has said that if France loses New Caledonia it will lose all its possessions in the Pacific, even Tahiti and Mururoa, where it tests its nuclear bombs. The *colons* are encouraged by the apparent right-wing determination to hold on to New Caledonia. Armed conflict has not yet finished.

The Kanaks will step up their struggle if the right-wing victory in France gives the settlers preferential treatment, as it was bound to do. The key Kanak threat is to strangle the capital. The Kanaks say that with the countryside in their hands they could cut off food, water and other supplies to Noumea.

Following the French elections in March 1986 the independence movement began a campaign to bring foreign pressure to bear against the French. John Peu, a prominent Kanak Liberation Front official, appealed in Australia for a total boycott of New Caledonia. In normal times many thousands of Australian tourists visit the island.

Despite this move, the Kanaks were conciliatory in internal politics. They promised that the new country, which they plan to rename Kanaky, would not expel people of other races.

Nicaragua—Sandanistas versus Contras

ANOTHER VIETNAM

Background

Of all the wars and civil wars taking place in the Americas, that in Nicaragua is the most dangerous because of the superpowers' involvement and the political and strategic importance which they both attach to it.

Because of this, Nicaragua, a poor country, has taken on a significance out of all proportion to its size, 130,000 km^2, and population, 3.2 million. Without an understanding of the background of recent Nicaraguan politics, it is not possible fully to understand a war which at times seems to have all the factors necessary to turn it into another Vietnam.

The corrupt Somoza family had ruled Nicaragua as a private fiefdom for decades. The cruel dictator Anastasio Somoza, who had enjoyed the support of successive US governments, realised to his alarm, in 1977, that the Sandanista National Liberation Front which opposed him had built up a guerrilla army which had popular support throughout the country.

The Sandanistas take their name from Augusto Sandino, an army general turned rebel who was assassinated in 1933. In 1979 the Americans saw that Somoza now had no credibility and "encouraged" him to flee the country, after a civil war in which 45,000 people died. (Somoza was later assassinated.)

The US welcomed the Sandanistas, particularly when they professed to be intent on bringing democracy to a land that had never known it. The Sandanistas quickly began to consolidate their rule, forming a junta in which the more radical left-wingers, such as Daniel Ortega, his brother Humberto and Tomas Borge, soon made life uncomfortable for the moderates who had supported the revolution.

The Sandanistas rule is élitist, even by Marxist–Leninist standards. The nine-man directory presides over an 80-member committee, and party membership cannot grow over 5,000. Nicaragua refused to join the large majority of non-aligned nations which condemned the 1979 Soviet invasion of Afghanistan. When military rule was imposed in Poland in 1981, the government issued a circular to the press instructing it to carry wire despatches on the crisis only from Tass and Cuba's news agency.

Nevertheless, in September 1981 President Carter assured Congress that the Sandanistas were not assisting left-wing insurgents in other countries, such as El Salvador. But 3 days after Mr. Reagan assumed office in November he cited Intelligence reports that Nicaraguan weapons *were* reaching El Salvador and cut off all American aid to Nicaragua.

At the same time he secretly approved funds for a covert war against the Sandanistas, to be run by the CIA and fought by *contra* (opposing) rebels with right-wing politics. From 1981 to 1983 the CIA ran the Contras operation through

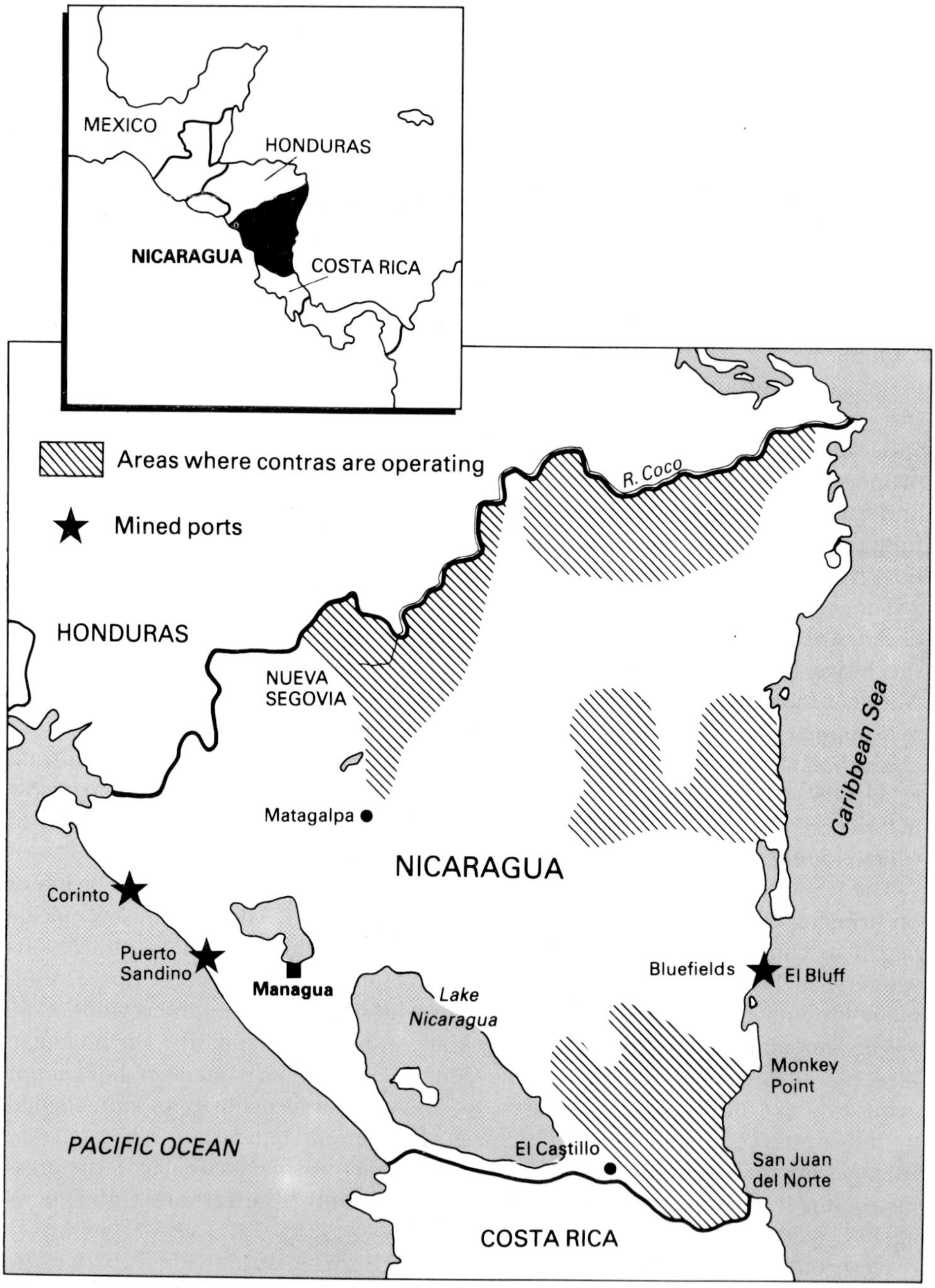

Nicaragua – Sandanistas versus Contras

agents of the Argentine Intelligence service, but dropped them because, as one CIA senior agent said, "They were taking 50 cents in the dollar for themselves".

The Sandanistas strengthened their relations with China and the Soviet Union and began to receive arms from the Soviet bloc, most of the supplies being channelled through Cuba.

President Reagan had over-estimated the capacity of the Sandanistas to destabilise Central America and his spokesmen at the UN have exaggerated the human-rights violations of the Sandanistas, but certain events in 1982 indicated that a Communist conspiracy was taking place. In February the Miskito Indian population of north-eastern Nicaragua was resettled by force in a number of camps 100 km further south. More than 30 Miskito villages were destroyed.

According to the regional officer of the UN High Commission for Refugees, more than 12,000 Miskitos were forced to walk for 4 days through mountains to tent villages in soaking jungles. The Miskitos were supposedly being moved from a "military zone" for their own protection, and the government story was that a Miskito leader from Honduras, Steadman Muller, was killing hundreds of Nicaraguan Miskitos in brutal raids. In fact, at least 12,000 Miskitos escaped from Nicaragua to seek refuge with Muller. The Sandanistas had two real reasons for moving the Moskito people from their lands. One was their resistance to being integrated into the Sandanist revolution. The other is that the Caribbean coast of Nicaragua is strategic territory for the Sandanistas; they have built several air bases and a port there and a military road links the Caribbean coast with the west of the country. The bulk of Cuba's 1,500 military advisers in Nicaragua are in the Caribbean area, which is intended to be Cuba's bridgehead on mainland America. The Miskitos were in the way.

The US Congress banned funds for overthrowing the Sandanistas and encouraged a diplomatic peace effort by other Latin American countries. Early in 1983 the Contadora peace initiative was launched; it was named for the island off the coast of Panama where representatives of Venezuela, Colombia, Mexico and Panama first met to begin the process. President Reagan declared his support for the Contadora effort, but at the same time authorised his Chiefs of Staff to begin a series of military manoeuvres in Honduras that were intended to intimidate the Sandanistas. In November 1983 Congress took Mr. Reagan's warnings seriously and approved $24 million in aid for the "Contra" rebel army of Nicaraguan exiles that had been trained with help from the CIA.

During 1984 it was disclosed that the CIA had been behind the mining of Nicaraguan harbours; several cargo ships from foreign countries were damaged. The CIA also issued an instruction manual to the rebels in which they were told how to assassinate civilian officials of the Sandanista government. Congress reacted by suspending further aid to the Contras. President Reagan stated publicly that he wanted to see the Sandanistas removed from office because they were a threat to stability and to democracy in Central America, and called on Congress to resume military aid to the Contras. His administration spoke for the first time of sending American troops to Nicaragua if all other anti-Sandanista efforts failed.

The President's advisers constantly point out that were it not for the Sandanistas the rebellion in El Salvador and the current threat to Honduras, Guatemala and Costa Rica would be manageable. Certainly the Sandanistas have publicly declared their desire to see revolution spread across Central America. When the Sandanista

National Liberation Front published its first detailed statement of goals in 1969, its manifesto called for "A struggle for the true union of the Central American peoples within one country, beginning with support for national liberation movements in neighbouring states". The Sandanist motto is *Oatria libre or morir*—A free country or death.

Despite the totalitarian and revolutionary nature of Sandanista politics, the party did hold reasonably democratic elections in November 1984 and in a 75% turnout won 63% of the votes. After this overwhelming success the rule of the nine-member junta under Daniel Ortega was confirmed. All but two of them—Tomas Borge and Victor Irado—are young; Ortega is 40. Youthful leaders have a great political advantage in a country where half the population is under 20.

Behind the government lies the standard apparatus for the enforcement of the revolution—the neighbourhood Sandanist "defence committees", "the popular militia" and the Sandanist Popular Army. As Tomas Borge observed to a Western journalist in 1982: "In Nicaragua the armed forces have the power—and the armed forces are revolutionary".

The result of the elections did not diminish the Contras' hostility. The term *contras* covers several rebel guerrilla groups. The largest, the Nicaraguan Democratic Force (NDF), is about 9,000-strong and led by Colonel Enrique Bermudez, a former Somoza officer. The NDF operates in the north of Nicaragua and from over the Honduras border. The main southern group is the Democratic Revolutionary Alliance (ARDE) which had about 3,000 fighters before 1985, led by the swashbuckling "Commander Zero"—Eden Pastora Gomez, a former Sandanista. Smaller guerrilla groups operate independently of these two major parties, and there are about 15,000 Contras in all. To help the Contras the American Army maintains 1,200 military advisers at Palmerola in Honduras. They have kept well clear of the fighting, which killed 5,000 people and wounded 6,000 during 1984.

The War in 1985

The Opposing Forces

NICARAGUAN ARMED SERVICES

The Sandanista Army has 60,000 regulars and 120,000 reservists, of which 30,000 are on duty at any one time. There is no officer corps and no general staff, but Humberto Ortega, as nominal commander-in-chief, has divided the country into seven military regions for ease of command and quick response.

The army comprises: 1 motorised infantry brigade, 3 armoured battalions, 12 infantry battalions, 3 commando (or "Irregular Warfare") battalions, a field artillery group of 3 battalions, 1 engineer battalion and 1 anti-aircraft group of 9 batteries.

The troops are equipped with M-16 and AK-47 (Kalashnikov) rifles and the army has some mortars, multiple rocket launchers, anti-aircraft missiles and Sa-7 launchers. The French have sold some shoulder-fired rocket-launchers. The armour inventory increased greatly in 1985, to 350 tanks, mostly Russian T-54 and T-55, and other armoured vehicles, mainly APCs. The gun mainly in use in 1985 was the 152 mm firing 90 lb shells a distance of 16 miles. Soviet military shipments to Nicaragua reached 18,000 tons in 1985, a 20-fold increase since 1981.

The air force of 1,500 men has only 30 combat aircraft, but these include some Soviet Mi-24 Hind D helicopter gunships, which give the Sandanista forces a great advantage in jungle warfare.

The navy has 13 small ships, including a frigate, but only 300 naval personnel. The navy's chief function is to patrol Nicaragua's two coasts, on the Pacific and the Caribbean.

Active military service applies only to men between 17 and 25, who are conscripted for 2 years, but draftees as young as 12 are willing to die for the Sandanista cause. Increasing numbers of middle-class parents are sending their sons out of the country to evade military service.

CONTRA GUERRILLAS

While well supplied in some ways, the Contras also have primitive equipment; for instance, new FDN recruits are armed only with World War I vintage Mauser bolt-action rifles, given by the CIA in 1982. Trained guerrillas have either Belgian FAL or Spanish-made G3 rifles or the West German G-3 automatic assault rifles. They have plentiful stocks of American M-15 landmines and dynamite and anti-aircraft missiles. In addition, they have two ancient DC-3 (Dakota) aircraft for air-dropping supplies to patrols inside Nicaragua.

STRATEGY: TACTICS: OPERATIONS

Superpower strategy influenced all military activity throughout Central America during 1985, and especially that in Nicaragua. The Soviet strategic objective is to control Central America or to destabilise it to such an extent that should a major war occur the US can be threatened from the south. To achieve this objective the Soviet planners' strategy is to support leftist regimes by using Cuba as a surrogate. For all Latin American leftists, the Cuban Revolution is the event of the twentieth century and Fidel Castro and his comrades are the great heroes of the era. For the leading Sandanista Tomas Borge, "Fidel Castro is Sandino reborn".

The US strategic objective is wholly defensive—to support right-wing government and groups and thus stem the movement towards Communism. The American strategy is to strengthen the rightists with massive military assistance and to intimidate the leftists by displays of military might.

The tactics of the war are dictated by the disproportionate strength of the opposing forces, by the terrain and by the advice of the foreign advisers. The rebels are restricted to guerrilla warfare, and on the few occasions when they have used conventional tactics they have been beaten. Because the older Sandanistas were once guerrillas themselves, they understand the Contras' tactics.

The war is fought in jungle-covered mountains, river valleys and coastal swamps. In a good half of the country's area no fighting has taken place, though the Contras sometimes plant time bombs in government areas.

Early in 1985 the Americans began a war of nerves against Nicaragua. US navy vessels patrol off the Nicaraguan Pacific port of Corinto, watching cargo vessels bringing in military material. US SR-71 spy planes fly over Nicaraguan territory daily. General Gordon, commanding the Panama-based Southern Command and responsible for all military operations in Central and Southern America, makes conspicuous and well-publicised surprise visits to Costa Rica, near Nicaragua's southern border. Gordon, who is known to be a hawk, favours "immediate direct action".

Apparently in direct response to the more aggressive US policy, Soviet bloc involvement increased markedly during the first half of 1985. By May there were between 6,000 and 7,500 Cuban advisers, in addition to several hundred from other Communist bloc countries. On Soviet advice, the Sandanistas set up a "directorate of state security" staffed by foreign advisers—400 Cubans, 70 Russians, 50 East Germans and 25 Bulgarians.

There is evidence that Nicaragua has a Soviet-financed and East European–staffed training centre for international terrorists attended by PLO fighters (from the Abu Musa Syrian-controlled faction), the Italian Red Brigades, the Montoneros of Paraguay and the Red Army Faction (successors to the Baader-Meinhof Gang).

On Cuban advice, the Ministry of Education sent a circular to schools calling for 2 hours a week of "patriotic education". In response the Roman Catholic hierarchy published an open letter in the only opposition newspaper, *La Prensa*, protesting against a programme which was "oriented to the installation in children's minds of a partisan mentality which encourages class hatred, violence, militarism and the warrior spirit".

Also on Soviet-Cuban advice, Ortega and his colleagues ordered the army to move 50,000 peasants from five northern provinces—in addition to the already moved Miskitos. This is part of a systematic plan to deprive the Contras of a popular base that provides food, refuge and recruits. By forcibly driving out the peasants for resettlement further south, the army was giving itself a free-fire zone. Here it is now using the Soviet Mi-24 Hind helicopter gunships. With rockets and rapid-fire guns, the helicopter crews can wage war against the Contras without the danger of hitting civilians. The army destroys the homes of peasant farmers who are moved out so that Contras cannot shelter there.

With Cuban and East European help, the Nicaraguan Army has trained some tough counter-insurgency units, the "Irregular Warfare Battalions", who go out on patrol for weeks at a time, living off the land. These units have ended any Contra hope of conquering enough land with a big enough town in which to set up an alternative government which Washington could recognise. The nearest the Contras have operated to Managua is Matagalpa, 60 miles north of the capital.

In May 1985 President Reagan, worried about the Sandanistas' success in containing the Contras, proposed to send the Contras massive aid. Congress, however, not only blocked all military assistance but food and medicine as well. In the face of this reverse, the President vowed never to abandon the Contras and ordered a trade embargo against Nicaragua.

Soon after this, President Ortega made a major tactical mistake by visiting the Soviet bloc; this tacit admission of friendship with the Communist countries greatly softened Congressional opposition to President Reagan's policy. Ortega also likened Reagan to Hitler, a comparison which did nothing to win him support, even among Reagan's enemies. The President's Contra-aid proposals were approved.

The Contra NDF's command centre and main camp is in the hills above the Coco River, for part of its course the border between Honduras and Nicaragua. In addition to quarters for 4,000 men, the camp has a hospital, training schools, long-range radio facilities and well-stocked supply dumps. The border here is ill-defined and whether the base is in Honduran territory is debatable. This gives it some political protection, but in any case it is well guarded against air attack.

During 1985 the Contras changed their tactics and attacked soft targets—villages without protection, state-owned farms, government health centres, agricultural co-operatives, and any meeting that could be called "government sponsored". The aim was to bring administrative collapse, which they achieved when they blew up four electrical towers in Jinotega, throwing that town and nearby villages into darkness for 3 days.

Priests, among others, confirm charges that Contra guerrillas are often guilty of torture in their efforts to obtain information about army movements or simply because they suspect that people of a certain village have reported their own movements to the authorities. The Sandanista forces are equally brutal.

However, late in 1985 the government began to operate a peaceful public relations campaign to encourage Contra fighters to return to civilian life. For instance, in the town of Pantasma, Jinotega, Sandanista agents made house calls to 160 peasant families believed to have relatives fighting with the Contra and offered them amnesty papers. Within a few months 60 Contras, who said they were forcibly recruited, responded by returning home. Sandanista representatives hold public ceremonies to celebrate such homecomings.

Twice during 1985 the Contras vainly tried to capture the army base at Bluefields on the Caribbean coast. On another occasion they attempted a conventional movement against an army base, but were trapped in a valley by infantry and finished off by artillery fire.

At Mulukuku military school, in the rain forest of central Zelaya 250 miles north-east of Managua, Sandanista army recruits often get their first taste of combat before they complete their course. From this base and elsewhere, the army occasionally sends a column in an attempt to overwhelm a Contra stronghold, though this has yet to succeed; the Contras do not stay to fight. The army's ambushes are more successful, mainly because they use their version of the Green Berets, an élite battalion of about 600 trained in anti-guerrilla warfare by Korean and Vietnamese instructors.

The Sandanistas experimented in 1985 with "hot pursuit" tactics—pursuing Contras into Honduras. Should the Nicaraguans enter Honduras too often, President Roberto Suazo could invoke the Rio Treaty, which states that an attack on one American country is an attack on all. The US and other countries in Latin America might then be obliged to come to the help of Honduras.

Without question, the Contra campaign is having a severe effect on the Sandanistas' ability to run Nicaragua; the fighting drains off half the nation's budget. But no matter how much support the US provides to the Contras, they have no chance of defeating the Sandanistas. To do so they would have to capture Managua and induce the army to turn against the Sandanistas. The regime can only be brought down by the open intervention of American forces.

In the meantime, the Contras face many problems. Pastora Gomez's ARDE group in the south is not only estranged from the FDN but is badly split internally. Gomez, the charismatic leader of 3,000 men a year ago, has been reduced to door-to-door fund-raising in San Jose, Costa Rica's capital. According to some reports, his followers have sold guns for food. In April 1984, Pastora's fighters captured the fishing port of San Juan del Norte after 4 days of fighting; they would not be capable of such a feat now.

Contra morale is low and units are held together more by fear of their fate should the Sandanistas defeat them than by their belief in the democratic cause. Leadership on both sides has been incompetent and the large numbers of foreign advisers and instructors, while greatly improving the competence of the rank and file, have done little to improve leadership. Short of American military intervention or a radical change in US policy, the war could drag on for years.

Much depends, for the Americans, on what use is made of Punta Huete airfield, which is the Central America's longest military runway. If Soviet and Cuban aircraft use it, the Americans might become alarmed.

Alarm and Desperation

President Reagan was certainly alarmed in 1986 when he asked Congress for $100 million more in aid. Even with that money the Sandanistas were confident that the Contras were no longer a serious threat. In March 1986 they sent crack units into Honduras to raid rebel camps. They killed 350 rebels and destroyed the main Contra training base near the Honduran town of Las Trojes.

This brought the Sandanista units into direct conflict with the Honduran army, which killed a few score Nicaraguans. A British diplomat reported that the Sandanistas had made about 200 incursions into Honduran border territory between September 1984 and March 1986—but the Honduran Government's official line was that "as there are no Contras operating from Honduras we cannot explain the Nicaraguan incursions".

Western diplomats in Managua said, in mid-1986 that the Sandanistas were close to dealing a strategic defeat to the Contras. "The government can deliver irreversible military damage to the Contras in 1986 if they get no military assistance", one of them said. This view was in stark contrast to that of March 1985 when the heavy Contra presence in northern Nicaragua was taking a direct economic toll. At that time the FDN had 4,000 troops there. According to the Sandanistas, no more than 50 remained a year later.

ARDE lost its main base camps near the Costa Rica border and, weakened by desertions, was reduced to occasional forays against government posts. The only major concentrations of Contras still inside Nicaragua—estimated by the Sandanistas at 1,500—were in the provinces of Boaco and Chontales. Apart from the military reverses, the Contras suffered, in 1986, from the Sandanistas' better Intelligence gathering and better training. The Cuban instructors had done their job well.

It is not surprising that President Reagan reached such a stage of desperation in his efforts to get aid to the Contras that he warned Americans that Nicaraguan terrorists and subversives and their allies were "only two days' driving time from Harlingen, Texas". He finally won from Congress the money he wanted. He might yet commit American troops to direct involvement in Nicaragua.

Northern Ireland Terrorist War

NO COMPROMISE

Background

In modern times violence in Northern Ireland goes back to the late 1960s, with the greatest violence in the early 1970s. The basis of the confrontation is the demand by the nationalist organisation Sinn Fein for the union of Northern Ireland with Eire, the Irish Republic, and the insistence by the Protestant or Loyalist community that Northern Ireland remain part of the United Kingdom. Sinn Fein's military arm is the Provisional Irish Republican Army, the "Provos", though in practice they have largely the same command. The Provos wage a terrorist war against the Protestant "defence" groups, some of which are also terrorist. The British Army's overall objective is to protect both sides in the endless dispute, while fighting the terrorists.

Since 1977 the British Army's responsibility has not been with law and order. From that year maintaining the peace has been the responsibility of the Royal Ulster Constabulary (RUC). Any operations which the army undertakes are carried out in co-operation with the RUC and at its request. In 1972, the peak year for terrorist violence, there were 10,630 shootings and 1,382 explosions.

The War in 1985

Fewer than 9,000 British regular army soldiers are on duty in Northern Ireland (22,000 in the early 1970s). The army comprises 8 infantry battalions and 9 battalions of the Ulster Defence Regiment, 6,500 men and women in all. Together they form 2 brigades, the 8th, with its HQ in Londonderry, and the 39th, in Lisburn, south of Belfast.

Six of the battalions are based in Londonderry, Ballykelly, Omagh Aldergrove, Holywood and Ballykinier. Two battalions are in West Belfast and South Armagh respectively. The latter place is known to the army as "bandit country", where the police need much army support.

In over 85% of the province the UDR alone is responsible to the RUC for support in maintaining law and order and defeating the terrorists. The UDR is the largest infantry regiment in the British Army, while the 7/10 Battalion UDR, operating in Belfast, is the biggest battalion in the British Army, with a strength of 1,360. The UDR is organised into 9 battalions in areas that conform to the divisional boundaries of the RUC. The army calls these areas Tactical Areas of Responsibility (TAOR). In all, there are 14 TAORs.

A UDR soldier may belong to one of four categories: regular, permanent cadre (PC), part-time (PT) or Greenfinches, the female corps, so known because of the colour of the women's uniform. Each battalion has 10 regular army officers and

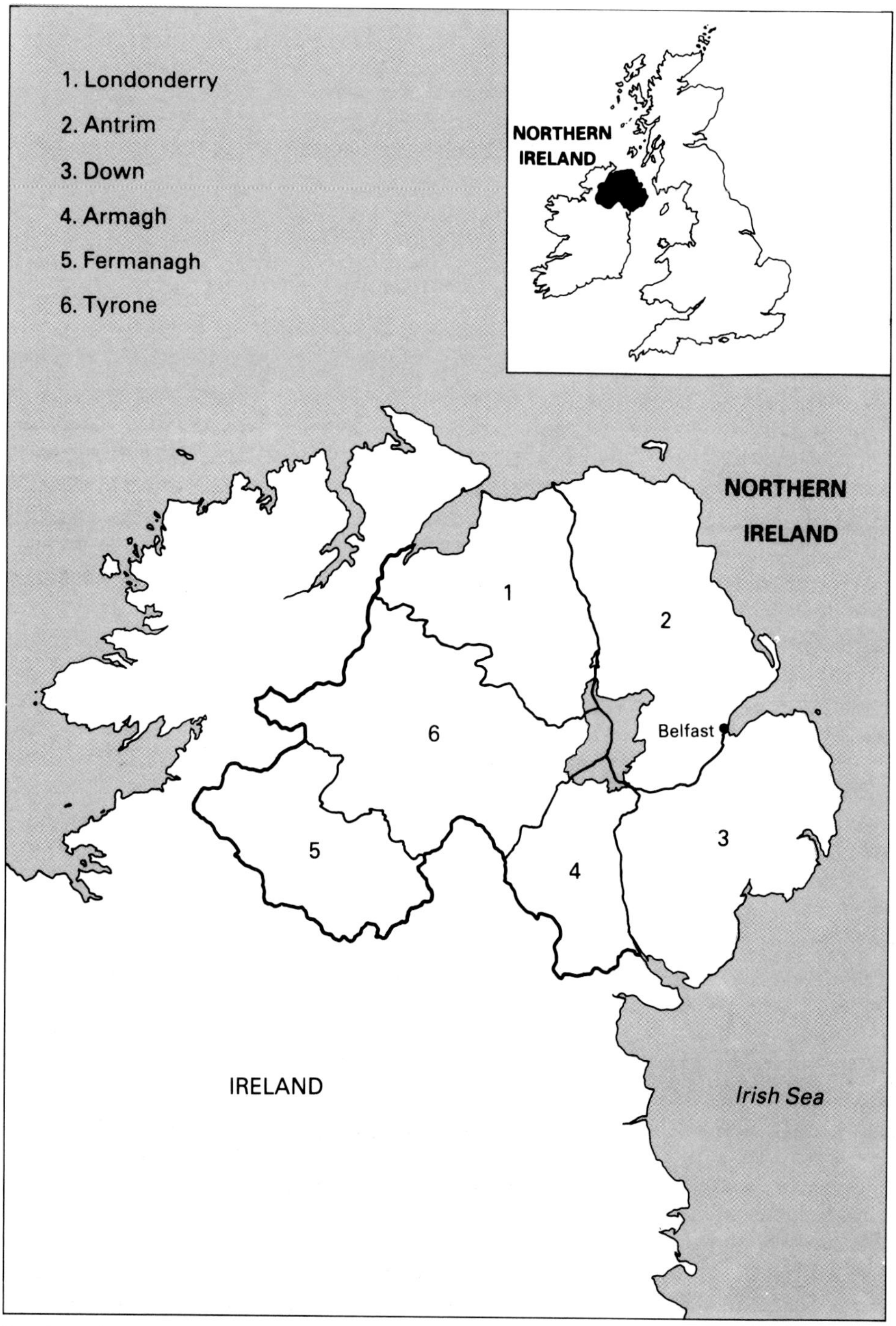

Northern Ireland Terrorist War

NCOs. The PC soldiers differ from the regulars only in that they live at home. The PT soldiers carry out their duties when they are not doing their daily civilian jobs, they are generally on duty one night in every three or four and most weekends. The army does not use them for covert operations, and they do not patrol in urban areas populated by extreme nationalists.

UDR soldiers use regular infantry weapons and will be issued with the new SA80. They travel in Land Rovers and wear normal infantry uniforms.

The UDR operates exclusively in Northern Ireland, and leaves Ulster only for training purposes. The men and women in its ranks do not return to the safety of barracks, as do the British soldiers, but to their own homes in communities where they are known. They run great risks and between 1970 and 1985 149 were killed; 121 of these deaths occurred while the soldiers were off duty. Many were killed at night by gunmen calling at their homes.

The continuous presence of UDR patrols and the refusal of its members to be intimidated has much to do with the reduction in terrorist activity. Because they are seen to be present at vehicle check points, searches and other activities they have a deterrent effect.

Military credit for the reduction in terrorist incidents is due not only to the more systematic use of the UDR but to improved Intelligence and greater use of British special forces such as the SAS—very often operating in plain clothes and unmarked cars.

The covert operations of the SAS and the regular field patrolling of the British Army battalions has also made it more difficult for the Provos and other terrorists to operate with confidence. It is known that the Provos are terrified of the SAS and its method of infiltration and intelligence gathering. Even more, the terrorists know that the SAS, when forced to shoot, do so to kill. In broad daylight in 1981 an SAS soldier in plain clothes, driving along an urban street, was stopped by four terrorists armed with machine-guns. With his pistol the SAS trooper killed three of the terrorists and wounded the fourth. Under his steady and disciplined counter-attack, the terrorists sprayed his car with bullets, but none hit him. That incident is still talked about in terrorist hide-outs.

Of the 13 active service unit men killed on operations between December 1984 and mid-1985, 9 died in SAS stakeouts, 2 died in incidents involving explosives and 1, an important activist, Sean McIlvenna, was shot by the RUC in Armagh. Another Provo, Kieran Fleming, was drowned while trying to flee from the SAS in Fermanagh. The Provos and INLA believe that the SAS intends to murder certain of their leaders.

In February 1985 a masked IRA gang ran into a dawn security trap on a hillside in the border town of Strabane. The security forces, in uniform, had the hillside staked out, awaiting the arrival of the gang. Three terrorists were killed and three rifles, two rocket-launchers and a number of projectiles were recovered. The killings indicate that the IRA is having serious internal security problems, with the police and the army becoming increasingly knowledgeable about specific terrorist operations.

While much money is raised for the IRA war-chest in the American Irish communities, the US authorities, during 1985, made it much more difficult for arms and explosives to leave the US. Simultaneously, the security forces in Northern Ireland have discovered several caches of terrorist supplies.

In 1985 detectives from the RUC undercover C19 racket-busting squad

A member of the Ulster Defence Regiment on duty in a fortified observation post near the border with Eire. (Courtesy *Daily Express*.)

Soldiers of the Ulster Defence Regiment on patrol in Londonderry. In their peacekeeping duties in Northern Ireland, these men lead a life of constant danger from snipers and bomb ambushes. (Courtesy *Daily Express*.)

discovered the Loyalist and IRA terrorists had carved Northern Ireland into spheres of financial influence to operate protection rackets, fraud and organised crime. This breakthrough showed that terrorist organisations were acquiring £5 million a year through such schemes.

It is not known how many terrorists belong to the IRA, which on paper has battalions and companies in the normal military fashion. The best Intelligence estimates put the number at 3,000. Many of these men and women live in the Irish Republic, and the number of "dedicated" active service terrorists at large in Northern Ireland at any one time is probably no more than 50. A small group can cause casualties and difficulties out of all proportion to their number, because they have the passive support of the Catholic community. Having made a terrorist strike, they find refuge in a Catholic home or across the border in Eire, where the British forces cannot pursue them.

The IRA's organisation is known and most of its leaders have been identified. The leader of Sinn Fein is Gerry Adams, who is also the member of Parliament for West Belfast—though he has never dared to appear at the House of Commons. The Provos' chief-of-staff is Martin McGuinness of Londonderry. One of the principal "active service units" commanders is Brendan McFarlane, who was chosen to lead the terror attack on the British mainland in 1985. Three members of his unit are known fugitive gunmen, Gerard McDonald, Kevin Barry Artt and Desnot McMelly. In 1976 McFarlane, 34, led a gun and bomb attack on a Protestant pub in the Shankhill Road of Belfast in which five people died. In 1983 he masterminded the escape of 37 fellow IRA prisoners from the Maze Prison and went on to lead the gang which kidnapped the Dublin millionaire Don Tidey.

The Provos are well supplied with modern weapons smuggled in from the US or Libya or, occasionally, stolen from British Army bases. The strategic objective of the terrorists is to maintain discord between the Northern Ireland Catholic and Protestant communities to prevent any agreement being reached on peaceful power-sharing government. In its strikes at the British Army, the Provos' tactics are those of the guerrilla—to kill and destroy by ambush and avoid pitched battle. In attacking the civil population, the Provos' tactics are outright terrorism—killings in churches and in private homes, bombs left in cars, shops and hotels. The bomb hidden in the Grand Hotel, Brighton, and blown up during the Conservative Party conference in October 1984, was one of the major terrorist attacks of the war. It was intended to kill the Prime Minister and members of the British cabinet, and it almost succeeded. The main terrorist aims are known to be the killing of members of the British royal family—Prince Charles is the main target—the Prime Minister, Secretary of State for Northern Ireland and several senior army officers.

The great advantage the terrorists have is that their options are practically unlimited and despite their losses in men—more than 200 were convicted by the courts in 1985—and the successes of the security forces the Provos' operations can be successful too.

For instance, in February 1985 the IRA made a mortar attack on the police post at Newry. One bomb landed on the post's canteen, killing 9 officers, during an evening meal break. Five weeks later a bomb killed a civilian and a constable outside the local courthouse.

In June, on a busy stretch of road south of Newry, four more officers were killed in a typical IRA attack. An RUC escort and a security van full of bank notes from

Dublin had just made a rendezvous. The armoured van had travelled from Dublin escorted by the Garda Siochana, the Irish police. Just as the Garda escort was withdrawing, 1,000 lb of explosives hidden in a nearby farm trailer were detonated. The blast ripped apart the RUC escort car with its four occupants. Only a few seconds before the explosion a minivan full of handicapped children had passed the murder spot.

Because the security forces have been remarkably successful in finding informers among the terrorists and their sympathisers, the Provos have adopted more callously systematic methods of punishment. The practice of "knee-capping"—firing a bullet into a suspect informer's knee—had long been an IRA punishment, but in 1985 entire groups of victims were being rounded up and driven to a hideaway for such treatment. Others have been attacked in their homes and beaten with baseball bats around the heads and legs. Many informers or suspects have been murdered.

Nevertheless, the number of shootings and bombings decreased in 1984 to 334 and 193 respectively. The steady fall is likely to continue, but there is always the risk of a terrorist coup being outstandingly successful—from the IRA's point of view. The organisation considers the murders of Earl Mountbatten, Airey Neave MP and soldiers at Warren Point, Northern Ireland, as among its "crowning achievements".

The Hillsborough Agreement signed by Britain and Eire late in 1985 was designed to reduce terrorism by increasing and improving co-operation between the two nations in Northern Ireland. It seemed to impartial foreign observers to be a sane and practical step. One benefit was that terrorists could not in future so readily find sanctuary in the Irish Republic.

However, a new kind of terrorism appeared as a result of the agreement. The Protestant majority of Northern Ireland, or that part of it which followed Dr. Ian Paisley, rejected the agreement and resorted to violence to show its displeasure. This violence reached a climax in March and April 1986 during the Loyalist (Protestant) "marching season". Following a ban on the march in Portadown, County Armagh, there was a pitched battle between police and protestant demonstrators; 39 policemen and 38 civilians were wounded.

The illegal Ulster Defence Association, the Protestant equivalent of the IRA, began to talk about armed conflict "to protect the Loyalist community from Irish domination." The armed conflict which the UDA and other paramilitary organisations had in mind was against the police.

Meanwhile the IRA also opposed the Hillsborough Agreement because it threatened the terrorists' freedom of action. It became clear during 1985–86 that the IRA had decided to make police and soldiers their main targets, rather than civilians. IRA attacks are meant not only to kill or intimidate the police; they are also designed to make constabulary stations inoperable. In 1984 46 police stations were attacked and in 1985, 33. Overall, by 1986 25 police stations had been destroyed or badly damaged by explosives or mortar bombs.

The most disturbing aspect of the war in mid-1986 was the scale and ferocity of Protestant attacks on the homes and families of the RUC. This was said to be a form of violent protest against the Hillsborough Agreement.

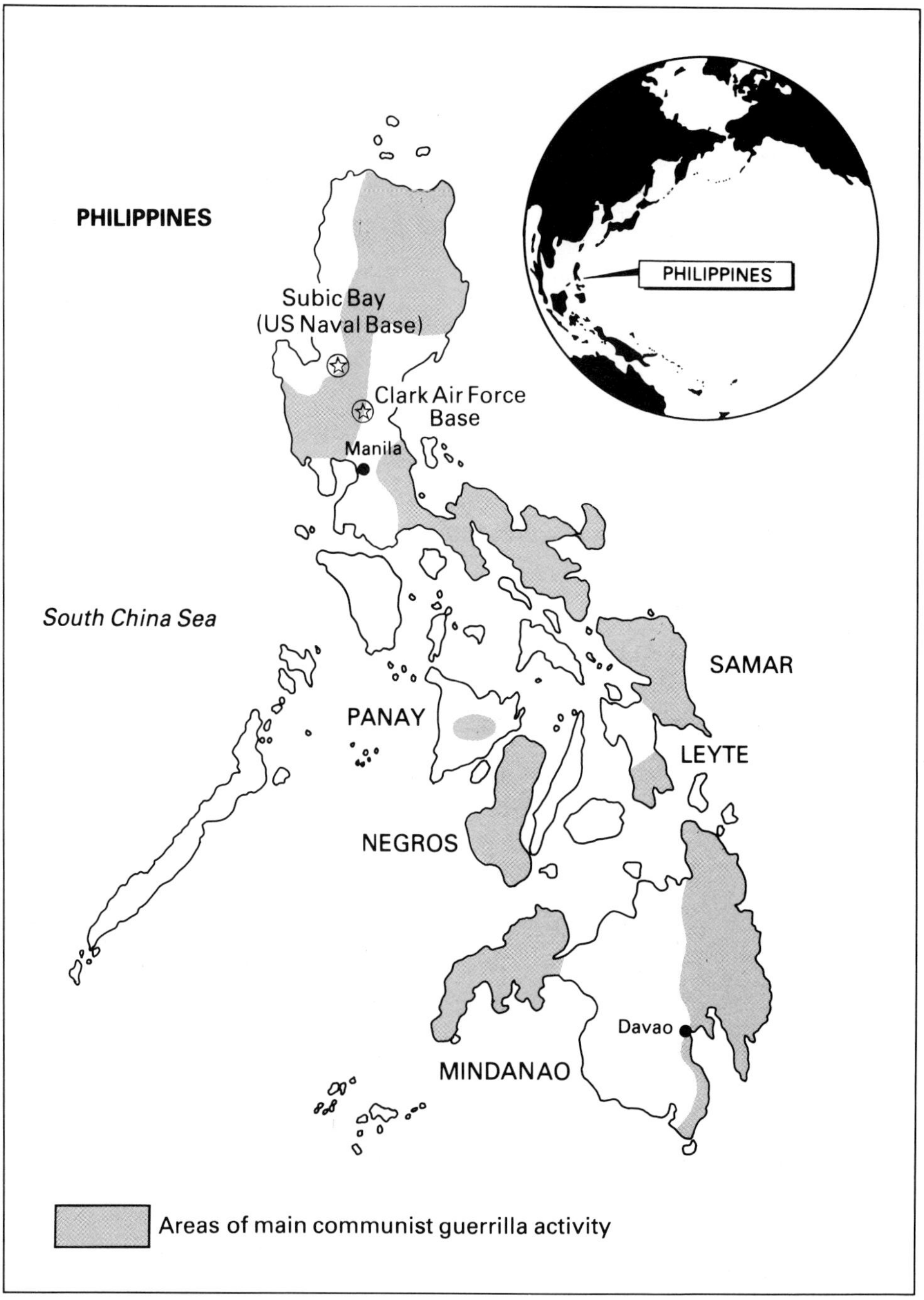

Philippine Guerrilla War

Philippine Guerrilla War

SOPHISTICATED STRATEGY

Background

During the period 1969-72 a new Maoist movement emerged in the Philippines, when a group of Communists, led by Jose Maria Sison, formed the Communist Party of the Philippines (CPP), together with a military wing, the New People's Army (NPA). In 1971 President Marcos, whose authoritarian parliamentary dictatorship had created great opposition, declared a state of emergency. He suspended some constitutional rights and detained 98 people after accusing NPA guerrillas of grenade attacks at election meetings. In July 1972 the government intercepted arms shipments from North Korea intended for the NPA. The army was sent in to northern Luzon to clean out NPA bases.

The NPA initially used Maoist "people's war" strategy and tactics. It appealed to peasants and to the newly emerging Philippine business class, which wants a smaller role for Chinese and other foreign businesses. Having whipped up militant nationalism, the CPP provoked urban demonstrations against Marcos, an advocate of foreign investment.

Marcos declared martial law on 21 September 1972 and cited the need to suppress the "state of rebellion" instigated by the NPA.

When, in 1976, nine NPA leaders were captured—they included Victor Corpus, NPA's chief of training—Marcos said: "I guess this just about eliminates the New People's Army". But a year later the NPA had joined with Muslim rebels seeking independence in Mindanao.

Just as the government seemed to be winning against the Muslim rebels, the NPA was hitting government targets. The government, frustrated by its inability to stop NPA activities, launched a 3-month anti-guerrilla sweep in Mindanao; 19 NPA leaders and 436 other Communists were killed. But 2 months later (29 September 1983) the NPA killed 39 soldiers and wounded 17 in Mindanao.

Marcos had set up what he called "groupings" in Mindanao; scattered villages were relocated into militarily supervised settlements to undercut local support for insurgents. He based this strategy on British success in Malaya in the period 1948-55, but it failed for Marcos because of army abuses against peasants. The NPA gained even more support from the peasants.

In April 1983 NPA changed its military tactics. Instead of small groups of three to nine guerrillas using hit-and-run tactics to seize arms from small bands of soldiers, NPA leaders decided to operate in bands of 200. Open, large-scale fighting gave them a regular supply of weapons. By 1984 NPA was operating in 60 of the Philippine's 73 provinces. In northern and central Luzon the army launched the biggest offensive ever against NPA rebels.

The War in 1985

NPA strategy, based on Mao's teaching and developed by Jose Sison, is highly sophisticated. It was to be in three parts, both in time-scale and policy. The first was blunt and brutal terrorism—hitting the government, army and police until they could stand no more. The second phase was military stand-off—there would be little military action by either side. Finally, by 1990, the final stage would be under way—strategic attack. The initiative is to pass to the CPP, the political wing of the NPA.

The NPA has been active on Negros since 1972; guerrillas first took to the hills in the aftermath of martial law in 1972. They found a ready welcome among the already uprooted peasantry forced from their lands by the master plantation owners and the government. Other main havens are the mountains of northern Luzon, Panay and southern Mindanao.

In 1985 the government appointed General Fidel Ramos as commander-in-chief of the armed forces—in succession to General Fabian Ver, who was implicated in the murder of Benigno Aquino, a noted opposition leader. Ramos announced that in the course of 3,500 NPA attacks 2,650 people had lost their lives and that 985 rebels had been killed. Independent sources confirm these figures. It was imperative, Ramos announced, that the "barbarians of the NPA" be stamped out.

In response, an NPA squad killed in a village street a retired general known and hated far and wide for his corruption and smuggling. They dumped his body in the square where they were greeted by the townspeople as heroes and liberators. A few days later an NPA sniper seriously wounded a general in broad daylight near his headquarters on the island of Mindanao. In both cases the guerrillas escaped.

Even the pro-government newspaper *Bulletin Today* acknowledges that the murder of security personnel by the NPA evokes approval among the people while the death of a rebel is greeted with sympathy.

The remarkable thing about the NPA's top leadership is its powers of self-regeneration. Time after time senior leaders have been captured, but the insurgent movement has not faltered. Nor has the motivation of the guerrilla rank and file. The NPA has no more than 20,000 fighters, but they can cause trouble to the security forces out of all proportion to their number.

The notoriously brutal attacks mounted by the military, especially in the rural areas, deprives them of any help from the ordinary people. Government troops are not psychologically trained for guerrilla warfare. Many soldiers suffer emotionally from the long separation from their families; some units have been on permanent anti-guerrilla duty for 2 years. Low pay and frequent long delays in handing out soldiers' wages make the troops susceptible to corruption, theft and smuggling. The constant danger of death causes wild, almost animal-like behaviour by some individual officers and soldiers.

Given these circumstances, the NPA's tactics—which are frequently bloody—easily win the favour of the village people. In talking with Western correspondents, the people refer to them as "helpful", "honest" and "goodhearted". In Davao, the NPA shows its other face—"its true face" according to the government. The guerrillas often kill, kidnap and blackmail to achieve their ends.

General Ramos, an intelligent man, evolved a counter political-military strategy, a "three-pronged push" he calls it. First, the military effectiveness of government troops would be increased following the US commitment to supply new arms and equipment to the government. Second, army violence against civilians would be stopped, offenders would be punished and the psychological training of the troops improved. Third, more attention would be paid to social and economic development, especially in Mindanao. The general's plans, ideal in principle, fall down to practice because of the inefficiency of many of his officers and the ignorance of his troops.

NPA exploits could be written into textbooks on how to run a guerrilla war. What happened during a weekend in May 1985 is a classic example. In the town of Isabela, on the island of Negros, officers at the headquarters of the 3rd Scout Rangers Company received information that guerrillas were gathering outside the town ready for a raid. A strong patrol was sent out on a Saturday evening to investigate, found nothing after 17 hours' searching and returned to report "All clear". The army post relaxed its guard.

About 120 guerrillas had come into town, without arms, as if joining the crowds of other people enjoying the usual Sunday fun. Their weapons entered town in market trucks. In the afternoon, during siesta time, the guerrillas hit the army camp and the town hall. Having killed the commander and about 20 soldiers, they got away with automatic rifles, machine guns, grenades, ammunition, radios and much else.

Army HQ in Manila at once sacked the colonel in charge of security on Negros and sent in 3,000 fresh troops of the 7th Infantry, under Colonel Moleta, backed by armoured cars. He at once posted detachments in every village—and succeeded only in further alienating the peasants. One farmer, seeing a helicopter on a search-and-destroy mission, became frightened and ran for his house; the helicopter gunner killed him. Strangers in the area are killed almost as a routine measure, and the soldiers loot any house they find empty on the grounds that it must belong to a guerrilla. In fact, few guerrillas live in houses. The army stays out of the mountains, where the guerrillas live, because it knows it cannot fight them on their own ground.

During 1985 the government formed a counter-insurgency force which is called the Civil Home Defence Force. It has induced planters in guerrilla areas to raise security squads from their own workforce. They are trained by the army, the planter gives them land for farming and they are supposed to be available on demand to fight guerrillas. The NPA—and many peasants—regard the Civil Home Defence Force members as traitors and several have been murdered.

Despite General Ramos's more enlightened approach to dealing with guerrillas, the army is also summarily brutal. In September 1985, after a gun battle in the northern Philippines, troops beheaded five captives including a Roman Catholic priest, Father Nilo Valerio, a guerrilla for 6 years. Scores of priests have taken to the hills with the guerrillas.

Late in 1985 the great foreign-owned sugar plantations became favourite targets of the rebels. With bomb attacks, they try to persuade the foreign owners, if not to withdraw entirely, then to make "regular financial contributions" to the NPA. In Mindanao the NPA, by orders, carries out four operations a week. The NPA exploit the country's chronic unemployment. On the island of Negros more than a quarter of a million unemployed sugar workers—of a total population of 2,000,000—struggle for survival and form a "natural" pool of recruits or helpers.

The NPA's main problem is finance. In 1985 it cost about 30,000 pesos ($1,700) to maintain a 100-fighter company in the field for a month. This is cheap compared with a conventional Western force, but the NPA has no outside source of funds. Ninety-nine per cent of NPA weapons are American-made as supplied to the Philippine Army and captured from soldiers. Most of the guerrillas' firearms are M-16 automatic rifles, together with M-1 rifles and carbines. The earlier North Korean source of weapons has dried up.

The Philippines Communist Party has no links with Moscow or Peking, and gets no military support from them. The financial backer of CPP is, surprisingly, the National Democratic Front, whose ideology is opposed to that of the CPP. But the NDF realised that the Marcos regime was a threat to national survival and that the CPP-NPA was the only force capable of preventing him from destroying the Philippines.

The Filipino armed forces are overwhelmingly superior to the NPA in every respect. Personnel total 104,000, of which 60,000 are in the army; the reserve army numbers 20,000, and 70,000 more have reserve commitments. The navy has 28,000 personnel with 202 ships, including 112 coastal patrol boats. Fifty of these are new and are intended largely for anti-guerrilla work. They have orders to prevent guerrillas or their couriers from travelling from one island to another. The air force has 16,800 personnel with 82 combat aircraft. The helicopter wing is the most important in the guerrilla war, and the Bell and UH-1H gunships have had some success against parties of NPA caught in the open, but this rarely happens.

In this country of so many islands and fragmented political groups, no one opposition group or force has enough strength to defeat the regime, but Manila intellectuals believe that in the face of so much unrest Marcos would again have to invoke martial law and that this would be followed by massive social repression. A general strike would follow martial law and if that did not bring down the government very large numbers of men would join the guerrillas in a large-scale civil war.

By 1987 the NPA aims to have 60,000 fighters in the field. By that time they will attack and briefly hold major provincial towns. Armed City Partisans (ACP) will be operating in most cities and towns. The partisans are already operating in Davao. The death squads call their work "salvaging".

For the American administration the guerrilla war in the Philippines is disturbing. The Philippines have long played a pivotal role in US defence plans. American military bases at Subic Bay and Clark Field are the most important in Asia and the largest in the world. Due west is the old US base at Cam Ranh bay on the coast of Vietnam and now used by the Soviet Union. The loss of the Philippines would be a tremendous blow to the US.

The American administration cannot afford to allow the Philippines to become a Communist-governed country, but it recognised the evils of the Marcos regime. It was forced to support Marcos in order to retain its bases. The Americans pinned their hopes to the large middle-class population of the Philippines which would not, in a free election, bring a Communist government to power. In fact, they elected Mrs. Corazon Aquino and Marcos fled the country. Nevertheless, according to an American Intelligence forcast, NPA "has a long life expectancy".

Sikh–Hindu War

VENGEANCE AND COUNTER-VENGEANCE

Background

When, in 1947, the Sikhs of the Punjab chose to remain part of India, they were demonstrating their ties with the newly emerging national conglomerate of faiths and languages. They had three demands—a Punjabi-speaking state, Chandigarh as the capital of the Punjab, a fair distribution of water and electricity.

None of these demands was ever fully met. Deeply aggrieved, the Sikhs fought for decades within the constitutional and political framework to redress their grievances. With a long reputation as fierce fighting men, in the 1980s they took the path of militancy and armed resistance to resolve a list of what had become 45 religious, political and economic demands.

Sikhs and Hindus fought a constant though undeclared civil war; conflict became more intense as motorcycle gangs of Sikhs armed with submachine-guns raided many Hindu villages. They drove into crowded village markets where they fired indiscriminately at fleeing shoppers; thousands of people were killed or wounded. The Sikh extremists also set fire to Hindu temples and lobbed hand grenades into others while Hindus were at prayer. Many of the raids were mounted from the Sikhs' Golden Temple at Amritsar. Vengeance-seeking Hindus hit back at Sikhs in Punjab and the neighbouring state of Haryana.

In addition, the Sikhs clashed with the security forces, sent in originally to guard the region's strategic border with Kashmir. In a single month, February 1984, 100 people were killed and 500 wounded.

Jarnail Singh Bhindranwale had become the spiritual and military leader of the militant Sikhs. Aged only 36, Bhindranwale preached an implacable fundamentalism and held unchangeable political views. Under his charismatic leadership, the Golden Temple's 72-acre complex became a formidable fortress.

The Sikh political party, the Akali Dal, tried again to reach an agreement with Indira Gandhi's government over their main grievances, but Sikh extremists sabotaged these talks. The Akali Dal leader, Harcharan Singh Longowal, called on his followers to prepare for an "all-out" war.

In the first week of June 1984 Mrs. Gandhi sent in her commandos to capture the Golden Temple and kill or drive out the Sikh extremists. In the resulting bloody battle an estimated 600 Sikhs and 80 soldiers were killed. The Sikhs' beloved Jarnail Singh Bhindranwale was one of the victims. The army continued its anti-Sikh operations in the Punjab; more than 400 Sikhs were killed during the latter half of 1984 and about 1,500 were imprisoned.

As a direct result of the Golden Temple battle, Mrs. Gandhi was assassinated on

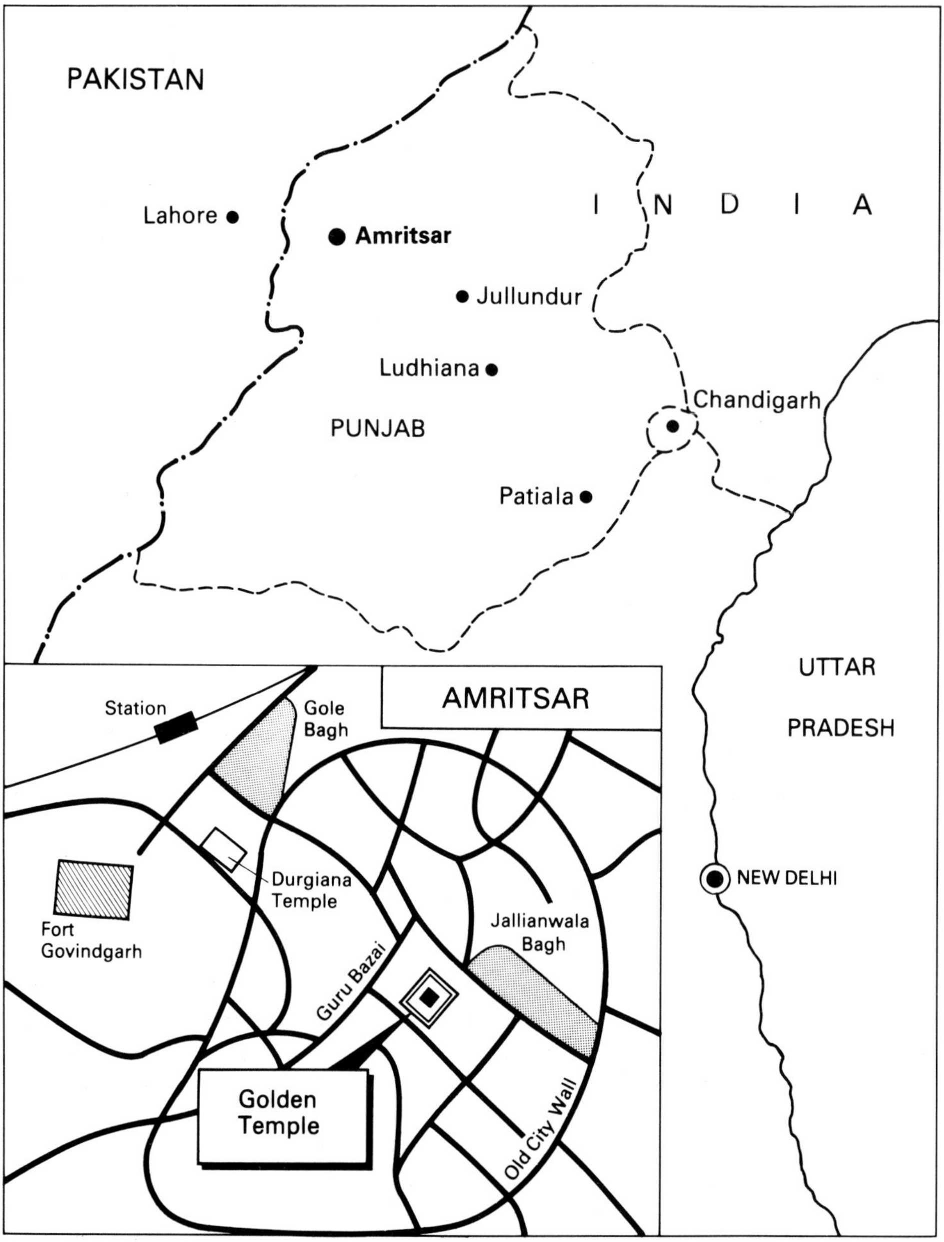

Sikh – Hindu War

31 October 1984 by two Sikh members of her bodyguard, acting under orders from a top-level Sikh planning committee. In turn, the Hindus sought vengeance against the Sikhs and 2,700 were murdered; many more were injured.

The War in 1985

In the Punjab any doubt that the name of the late Sikh militant leader Sant (Saint) Bhindranwale remains a potent factor in the state's politics was erased in May. His followers took their first step in wresting leadership of the Sikhs' Akali Dal movement from the moderates. Bhindranwale's father, Joginder Singh, aged 84, announced that he was now the leader in place of Harchand Singh Longowal, though Longowal disputed this.

Immediately after this change of leadership the Sikhs began a campaign of terrorism. On the first day 30 boobytrap bombs, many disguised as transistor radios, went off in trains, bus depots and parks in Delhi, as well as in towns of Haryana, Uttar Pradesh and Rajesthan. Many Hindus died in the bombings, which occurred at the places where Hindus had killed Sikhs.

Joginder Singh called on all Sikhs—there are 14 million in India and millions more abroad—to observe the week 1–7 June as "genocide week", an allusion to the army's attack on the Golden Temple and its later hunting down of extremists. He called on Sikhs to continue the "holy war" for more religious and political autonomy. He was addressing a convention at the Golden Temple, attended by more than 10,000 Sikhs, many of them youths wearing daggers, a practice declared illegal by the government. Another speaker urged Sikhs to learn the lessons of the army's anti-Sikh operations and prepare to defend their holy places in future.

Rajiv Gandhi, who had succeeded his mother as India's leader, brought in anti-terrorist laws which prescribe the death penalty for convicted terrorists. The Punjab was one of the first states to implement the new law. The governor of Punjab, Arjun Singh, a Sikh himself, said that the law was not intended to terrorise the Sikhs, but insisted that incitement to violence through fanatical religious songs would attract heavy punishment.

Gandhi's diplomacy in seeking co-operation between the government and moderate Sikhs suffered a severe setback late in August when Harchand Singh Longowal was assassinated. Much had depended on Longowal, the principal opponent of the vengeful Joginder Singh; he was the only identifiable representative of the moderate Sikh majority.

The elections in the Punjab in September brought into power many moderate men, but a Sikh "holy war", once called, has a momentum that is difficult to stop. The Indian Army high command had prepared, late in 1985, for "a war of indeterminate duration".

It was also evident that it would be fought outside India as well as inside. Militant Sikh underground groups, such as the Dal Khalsa, the Babbar Khalsa and the Dashmesh Regiment, expanded in Europe, Canada and the US.

The new Sikh strategy is to base their military leadership abroad, principally in Britain, where it is not vulnerable to Indian Army attack. Action within India is being left to many hundreds of cells with from 6 to 30 members, only one member of which will know the identity of a link member in another cell. Coded messages between the British HQ, sent by telex or courier, will activate the cells and give them their mission. Money from wealthy Sikhs abroad finances the operations in Punjab and other parts of India. Pakistani Intelligence agents have set up training camps for Sikh rebels in the border regions of Punjab state and the Jammu region. Camps are already established at Daska and Sialkot, in areas closed to the general public. One of the key organisers is Bhai Kanwar Singh of the Akali Federation.

During 1986 the cycle of vengeance and counter-vengeance appeared to be endless. In a very real sense the war spread beyond the borders of the sub-continent. Sikh terrorists were believed to have planted the bomb which blew an Air India Boeing 747 out of the sky and into the Atlantic Ocean with the loss of all 400 people aboard. Prime Minister Ghandi said that he feared a terrorist war in the Pungab would continue for years.

War in South-East Asia

The countries directly involved are Vietnam, Kampuchea (formerly Cambodia), China, Thailand and the Soviet Union.

A COMPLEX OF WARS

Background

Vietnam invaded Kampuchea in December 1978. Between the years 1975 and 1978, when Kampuchea was under the rule of the Khmer Rouge movement, the Vietnamese leaders Le Duan and Le Duc Tho were convinced that Kampuchea was becoming a springboard for a major attack by Vietnam's implacable enemy, China. They invaded Kampuchea to "protect their south-western flank from hostile influence".

For two reasons the Soviet Union backed Vietnam from the start. The first was that an alliance with Vietnam gave the Soviet leaders the use of the most strategically placed country in South-East Asia. The second was that China backed Kampuchea—and the Russians saw the Chinese as a future enemy in controlling the whole of the Orient.

Thailand became involved quite involuntarily because the Kampuchean resistance guerrillas—those fighting against the Vietnamese army of occupation—used Thai territory as their base. Also, hundreds of thousands of Kampucheans fled to Thailand as refugees. The US is indirectly involved in the war as a Kampuchean backer, though American support stops well short of military intervention.

The Association of Southeast Asian Nations—Thailand, Malaysia, Indonesia, Singapore, the Philippines and Brunei (ASEAN)—is an anti-Communist grouping which sympathises with Kampuchea's plight and accuses Vietnam of empire-building and of upsetting the region's military balance.

While Kampuchea has several supporters, Vietnam has only the Soviet Union, but its aid is considerable—an estimated $2 billion a year on military and technical assistance, partly in return for the use of the former huge, elaborate US naval base at Cam Ranh Bay.

The government of Kampuchea is a shaky coalition between the group led by Prince Norodom Sihanouk, the Khmer People's National Liberation Front, which is pro-Communist but anti-Soviet and anti-Vietnam, and the Khmer Rouge, the anti-Communists led by the infamous Pol Pot, whose men murdered Kampucheans in their millions during the period 1975-8.

Shortly before the Vietnamese invaded Kampuchea in 1978 they formed the Khmer People's Revolutionary Party. It had less than 100 members, Khmer communists who had spent the Khmer Rouge years in Vietnam or who had defected from the Khmer Rouge at various times. The Vietnamese hoped to use

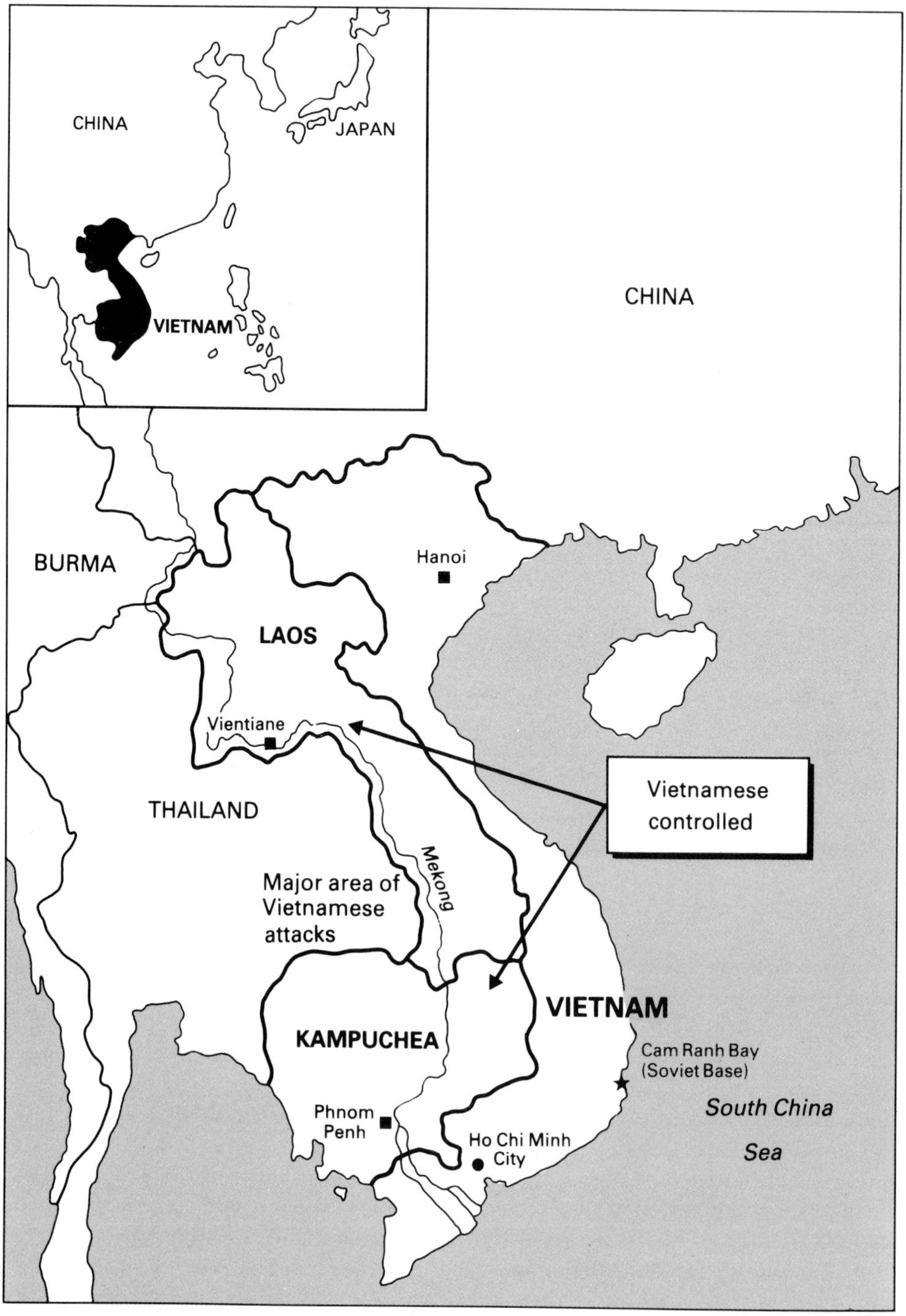

War in South-East Asia

this party as the basis for Communist government after the expected victory. Victory did not come because of the unexpected strength of the resistance movements, all of which built up guerrilla armies in the jungles and mountains.

In 1979 the Chinese Army invaded Vietnam "to teach the Vietnamese a lesson", as they said; the objective was to force the Vietnamese to withdraw their occupation army from Kampuchea. The Chinese high command was ill prepared for a war against an army that had been successfully fighting for years against the French, the Americans and the army of South Vietnam (which was incorporated into the national Vietamese Army after the US withdrawal in 1975).

The Chinese troops who then crossed the 190-mile Vietnam border were badly led, they had little air support and they were often lost in fog-covered mountains. The maps issued to the officers dated from the Manchu dynasty, which ended in 1972. At least 20,000 Chinese troops were killed and 80,000 wounded.

After this debacle the war in Kampuchea between the Vietnamese army of occupation and the resistance guerrillas settled into an attack and counter-attack conflict. It was not a genuine guerrilla warfare, because the Kampuchean resistance established fixed lines of defence and bases. When threatened by overwhelming Vietnamese force, they withdrew into Thailand, despite frequent Thai attempts to stop them and to evict them, or into the jungles.

During offensives before that of 1984-5 the Vietnamese always shelled the rebels' camps, driving scores of thousands of non-fighters into Thailand. The army then retreated from the border posts relatively quickly. The guerrillas always managed to rebuild their wrecked bases during the rainy season, beginning in May. When they did so, as many as 100,000 Khmer refugees would flood back to the Kampuchean side.

The War in 1985

Vietnamese strategy and tactics changed for the 1984-5 offensive. The Hanoi government ordered the army high command to eliminate resistance altogether. Two additional divisions were brought in, bringing the number of Vietnamese troops in the Thai border area to 60,000—160,000 in all in Kampuchea. The Vietnamese also conscripted tens of thousands of Kampucheans to build roads and lengthen airstrips in the region.

The offensive against the Khmer resistance strong-holds on the Thai-Kampuchean border was well planned and fiercely pressed. Rather than a concerted push against many enemy positions at once, the Vietnam high command opted for a series of strikes at one border "fortress" after another. Artillery was massed and opening up towards dawn the guns pounded a guerrilla base, which might have a garrison of perhaps 5,000. Vietnamese infantry advanced close behind the bursting shells and attacked the guerrillas in their trenches.

Son Kann's Khmer People's National Liberation Front lost all its bases. The formidable Khmer Rouge lost their mountain stronghold, Phnom Malai, in February. All that was left of guerrilla territory in Kampuchea was Tatum, in the north, held by supporters of the third member of the coalition, Prince Sihanouk.

Some people thought that the Vietnamese would not bother about Tatum. It was, in military terms, the least important of the bases, and Sihanouk's supporters are not very enthusiastic guerrillas. But Prince Sihanouk is not only president of

Khmer Rouge guerrillas, who are usually camera-shy, photographed with their grenade-launchers near their Phoum Thmei stronghold. Vietnamese troops captured part of the base in February 1985, but lost it again to these fierce jungle fighters.

"Democratic Kampuchea", he is also China's special friend, the most respected member of the coalition and he has a home in Peking. Vietnam waited until the very day that the Chinese President, Li Xiannian, was being received by the Thai king in Bangkok to attack Tatum. Then the Vietnamese Army shelled Tatum and sent in the infantry who used poisonous gas in their final assault on Tatum's clifftop trenches. Sihanouk ordered a "strategic withdrawal".

Two days later the prince arrived at the Nong Bua (Lotus Pond) Temple in the Thai town of Surin for the cremation of one of his generals killed during the campaign. The prince conceded that the Vietnamese victory was "very impressive".

Indeed it was, as the Vietnamese Army had needed only 16 weeks to secure its victory.

"We have lost our strongholds", Sihanouk observed, "but we control many villages". When Western correspondents pointed out that 50,000 Khmer guerrillas had been driven deeper into the interior of Kampuchea Sihanouk said: "We will see if the Vietnamese can maintain themselves in our strongholds".

During 1985 the Vietnamese Army command also evolved new tactics to deal with Kampuchean rebels based in Thailand. Scores of standing patrols of about 150 men each are sent to take up and hold positions at vital points on jungle tracks. These points might be track junctions, river crossings and mountain passes. The plan is to force the guerrillas along the few tracks left open where they can be attacked by the Soviet-made Hind helicopter gunships. Should the guerrillas decide on a pitched battle with one of the Vietnamese Army posts, it can call for rapid assistance from another post, possibly in the rear of the guerrillas. In one vulnerable area at Aranyaprathet the Vietnamese built an 8-mile long barbed wire fence and planted 10,000 mines and booby-traps to prevent guerrilla infiltration.

These tactics were among the many indications during 1985 that Vietnamese senior leadership was being more imaginative and that junior leaders were being given more authority to carry out minor operations on their own initiative.

Despite its military victory, Vietnam struggled throughout 1985 to extricate itself from its diplomatic isolation. The government indicated that its troops could be gradually withdrawn from Kampuchea in return for a simultaneous cut-off of Chinese aid for Khmer Rouge, denial of sanctuary for guerrillas and the irreversible exile of Pol Pot. Hanoi demonstrated a willingness to co-operate with the US by promising to return the remains of 26 US servicemen listed as missing in action in Vietnam. It would also assist the US in accounting for an estimated 2,464 other missing Americans.

Vietnamese Foreign Minister Nguyen Co Thach suggested negotiated power-sharing with Prince Norodom Sihanouk. The announcement that Vietnam would withdraw all its forces by 1990 does not mean that it is prepared to compromise on the primary issue—political control of Kampuchea through its intermediary, the Marxist Khmer People's Revolutionary Party.

Hanoi would be happy to see certain coalition-government leaders—namely Prince Sihanouk and to a lesser extent former premier Son Sann—given positions in the Kampuchean government so long as these are purely ornamental and do not dilute the dominance of the Vietnam-backed Kampuchean government. Hanoi has long been trying to woo the prince, apparently still popular among the ordinary Khmers, away from the coalition government.

Sihanouk has more belligerent ideas and has pleaded with the Chinese leaders to attack Vietnam. The Chinese Army had improved since 1979, and the officers, at least, are better educated. But Vietnam's army has improved too, and has had 6 years more fighting in Kampuchea. Chinese and Vietnamese troops not infrequently clash on the border. According to the Chinese, the Vietnamese are the aggressors; they often advance into Yunnan Province under artillery fire. In mid-1985 China claimed that the Vietnamese had fired 180,000 shells into Chinese territory in the previous 3 months.

The Chinese are not interested in an invasion of Vietnam, since it would involve some risk of Soviet retaliation—the Russians value their bases in Vietnam.

Cam Ranh Bay is really the Soviet's first true overseas base. New maintenance, communications and intelligence facilities have been built. During 1985 a squadron of 20 Mi-23 swing-wing fighters arrived at Cam Ranh, which also possesses a large airbase. The MiGs joined about 20 other Soviet warplanes already stationed at Cam Ranh. On any given day there is an average of 25 to 30 Soviet naval vessels, including about 5 submarines, anchored in the bay.

The Soviet Union is stepping up military aid to Vietnam with the training of a submarine arm for the Vietnamese Navy and the delivery of more tanks and armoured personnel carriers. The acquisition of submarines by Vietnam caused great concern in China and other South-East Asian mainland states such as Thailand which, while having no submarines, has become the first Asian customer to buy the advanced British lightweight Stingray torpedo.

The Soviet Union and Vietnam have had large-scale joint military exercises in Vietnam; in the most recent exercise the aircraft carrier *Kiev* and the assault ship *Ivan Rogov* took part. The Vietnamese ships which took part included the many small missile attack boats and frigates provided by the Russians in 1979. The growth of Soviet military strength in Vietnam worries the ASEAN countries which have long been concerned about Vietnam's conflict in Kampuchea and its other military objectives.

At the same time, friendship between China and Vietnam would be bad news for the ASEAN countries. Two of the ASEAN countries, Indonesia and Malaysia, believe that China is a greater threat to the stability of South-East Asia than Vietnam is, and are convinced that Vietnam helps to keep the Chinese in check. The others believe that China keeps Vietnam in check.

Throughout 1985 the ASEAN nations worked on a total reappraisal of strategy in the light of the Vietnamese superiority over the Kampuchean guerrillas and the build-up of Soviet military strength.

Whatever the ASEAN nations might do, the character of the conflict is changing. Throughout 1985 China, the US and Thailand advised and urged the Kampuchean rebels to become real guerrillas, that is to regroup into small, mobile units that could strike swiftly and vanish, rather than relying on static defence. The overwhelming Vietnamese offensive finally convinced the guerrilla leaders that they could not fight and win the war from fixed camps.

If Thailand is prepared to support hundreds of thousands of refugees—the families of the guerrillas—there is every indication that in 1986-7 the war in South-East Asia will become more ruthless as the Kampuchean resistance take up genuine guerrilla tactics.

Late in 1985 there were many signs that the Kampuchean Resistance had become more professional and disciplined. At the same time, it seemed that the

Vietnamese high command was moving its best units to the frontier with China and transferring units of poorer quality to Vietnam.

Some Western observers were permitted to accompany a KPNLF (Khmer Rouge) unit, the 215th (Commando) Battalion, on a raid against the Vietnamese. Its mission was to destroy a bridge on Route Six, a main supply line between Phnom Penh and Sisophon. The correspondents were unanimous in reporting that the job was well done, although the fighting force of 160 men were not all armed. Only 112 had weapons, mostly outdated Chinese type 56 carbines and assault rifles and US M16AI rifles. Back-up weapons were RPG-2s and RPG-7s, as well as 60-mm mortars. Ammunition was also in limited supply.

The Vietnamese ambushed a KPNLF platoon by night and the Kampucheans' reaction demonstrated good training and discipline. The leading section, which was pinned down, made no attempt to get out, but gave covering fire for the withdrawal of the other two sections which got into position to outflank the Vietnamese and cover the withdrawal of the first section. By fire and movement the three sections disengaged with the loss of only one man.

The entire force moved on to its objective, the bridge, after a 15-km trudge through water. The bridge was blown up and three Vietnamese posts were attacked. As the bridge had been undefended, the assault leader calculated that the Vietnamese would send a force to investigate the explosion, so he set an ambush. The Vietnamese soldiers walked straight into it and were hard hit.

Battalion 215 then withdrew and returned to base, having covered at least 180 km in the 6-day operation. Such a mission would not have been possible without the support of the civil population, an indication that the Vietnamese have made no progress with their campaign to win the Kampucheans' sympathy.

Capture by Colonisation

Vietnam's stated intention to pull all its troops out of Kampuchea by 1990 is not implausible. Following the Vietnamese invasion in 1978 large numbers of Vietnamese families have been settled on Kampuchean land. By 1990 at least 800,000 Vietnamese or people linked to Vietnam by marriage will be living in Kampuchea. Vietnam might then feel that its occupation troops can be brought home.

Most colonisation is taking place in Kampuchea's south-east, along the Vietnamese border. This is no coincidence. The plan is to form a buffer zone against possible future attacks by an independent Kampuchea. As Kampuchea has a population of only 7 million against Vietnam's 40 million such attacks seem unlikely but the Vietnamese leaders are obsessed by defence considerations.

A future Kampuchea with a large and expanding permanent Vietnamese minority does not appeal to Kampucheans, even if the occupying army were to withdraw. Son Sann, prime minister of the resistance coalition, complained in 1986 that "soon nobody in the country will know who is Vietnamese and who is Kampuchean". This is precisely Vietnamese strategy. Vietnamese influence will grow through colonisation and as a result of division among Vietnam's enemies. The third Indochina war (the first being against the French and the second against the Americans) will end with a victory for the Vietnamese communists, this time as much by stealth as by steel.

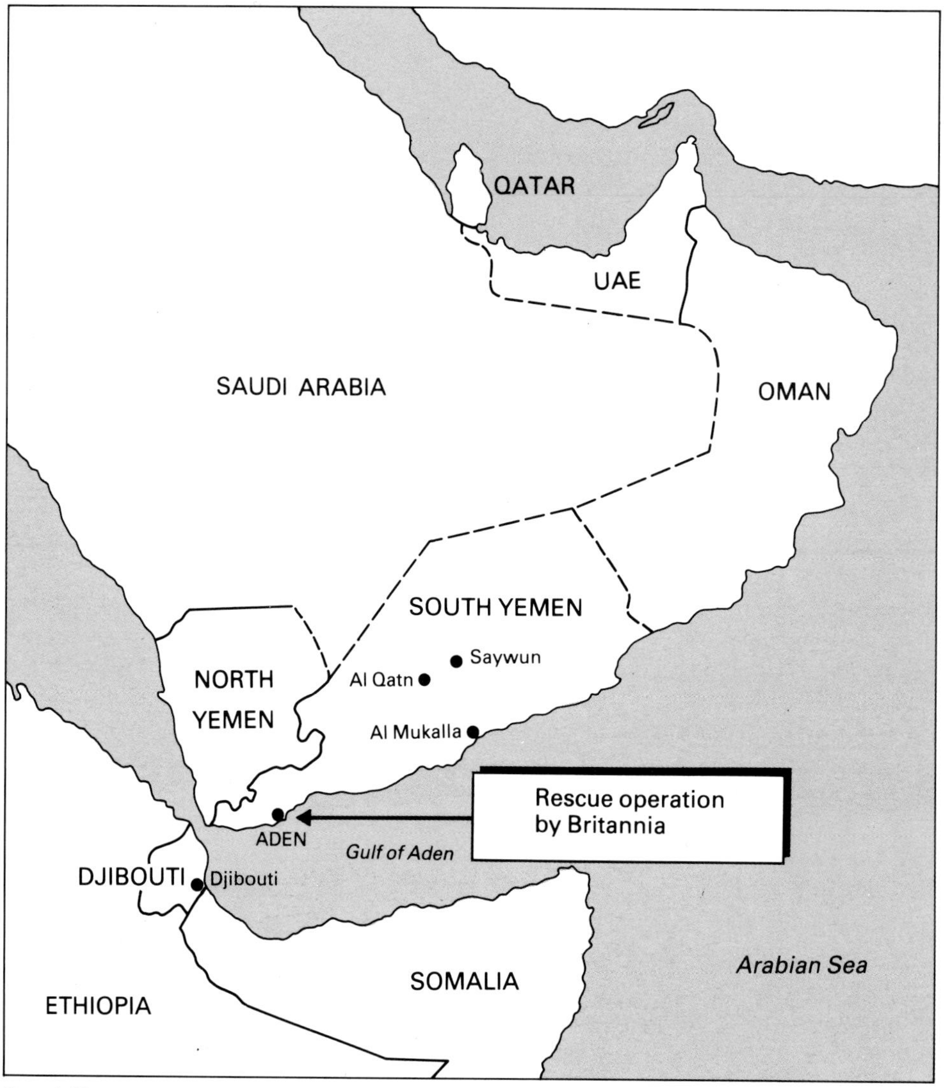
QATAR
UAE
SAUDI ARABIA
OMAN
SOUTH YEMEN
Saywun
Al Qatn
NORTH YEMEN
Al Mukalla
Rescue operation by Britannia
ADEN
Gulf of Aden
DJIBOUTI
Djibouti
SOMALIA
Arabian Sea
ETHIOPIA

South Yemen

South Yemen Civil War

COMRADES FALL OUT

Background

The fierce fighting in the People's Democratic Republic of Yemen in January 1986 between forces loyal to President Ali Nasser Muhammed and his rival Abdul Fattar Ismail was a surprise, but a confrontation between the two factions was to be expected sooner or later. On an ideological basis, the struggle pitted the pragmatic Marxist Muhammad against the more hardline anti-West Ismail.

Better known as South Yemen, and a Marxist state despite its "democratic" label, the YPDR includes the former British colony of Aden. The Yemenis fought a war of independence against Britain, which withdrew in 1967. A country of 300,000 m^2, the YPDR has a population of 2.25 million, most of them with tribal affiliation.

Violence was endemic from the beginning. In February 1967 a former minister, Said Muhammad, was murdered and in 1970 Faisal Sha'abi, a former prime minister, was shot "while attempting to escape from his prison". In March 1970, 40 opposition leaders were invited to attend discussions with the government; all were shot to death. In 1978 the chairman of the presidential council, Salem Ali Rubayyi, and two of his ministers were executed in a coup. The country was then controlled by Abdul Fattah Ismail, leader of the Yemeni Socialist Party (YSP).

During his term as president between October 1978 and April 1980 Ismail alienated his country's neighbours, North Yemen, Oman and Saudi Arabia, by encouraging rebels in border areas and following a hardline pro-socialist foreign policy. His government signed a 20-year friendship treaty with the Soviet Union in 1979; this allowed the Russians to station up to 18,000 men in South Yemen.

The only Arab country that specifically calls itself Marxist, South Yemen quickly strengthened its links with the Soviet Union. The Russians built a military base at Aden and an advanced electronic listening post on the Yemeni island of Socotra, 300 miles offshore.

Ismail's policies brought economic hardship to his country, partly because foreign investment was reduced to a trickle. He resigned "for health reasons" in April 1980 and was replaced by Nasser Muhammad. The Russians had encouraged the change because they preferred a more pragmatic leader in South Yemen.

Nevertheless, Moscow wanted to keep Ismail under its wing as a loyal friend whose continuing influence in South Yemen could prove useful in the future; he was popular with large sections of the 15,000 paramilitary People's Militia and the 30,000 public security force, all trained by East German instructors. He was sent to live in exile in Moscow.

President Muhammad opened up South Yemen to investment by Western and Japanese companies and increased trade with non-Communist countries. His government obtained substantial amounts of aid from the Gulf countries and from international development funds. He also improved his country's relations with North Yemen. In July 1983 Aden and Riyadh raised their diplomatic relations to ambassadorial status. Also South Yemen and Oman laid their long-standing dispute to rest and established high level diplomatic ties.

Throughout this time Muhammad scrupulously consulted Moscow before all of his important policy decisions, even about the extensive purges of the country's armed forces leadership and the government bureaucracy.

However, he was unable to remove several ideological hardliners from the top leadership, including Defence Minister Qassem, Deputy Prime Minister Ali Badeeb, Construction Minister Fadil Abdullah and Local Government Minister Ali Al Beedh. These and other radicals in the party ensured that Ismail could eventually return to Aden as a force to be reckoned with.

Late in 1983 Ali Antar, former Minister of Defence and Chief of Staff—and until the 1986 civil war, Vice President of the country—visited Moscow where he tried to enrol Abdul Ismail in his efforts to overthrow President Muhammad.

The stage for confrontation was set in February 1985 when Ismal did return to Aden after nearly 5 years of exile. Soon after his return the central committee of the YSP appointed him as its secretary. A session of the Supreme People's Council in October confirmed him in that position. The YSP's Politburo was expanded from 13 to 16 members and its central committee membership grew from 47 to 77. Many of the new members on both bodies were supporters of Ismail and Antar.

The War in 1986

The fighting began on 13 January 1986 when Aden radio announced that security forces had foiled a coup attempt that involved a planned assassination of the president. It said that unnamed "imperialist and reactionary circles" were behind the plot and that four ringleaders had been executed, including Ali Antar.

But the armed forces split almost exactly in half in their support for one or other of the rival leaders. Both groups of combatants had access to the equipment which the Soviet Union had supplied to South Yemen. It included 450 tanks (T-54s, T-55s and T-62s), 400 other armoured vehicles, 103 combat aircraft and an array of SA missiles, Scuds and FROGS. With such equipment and large stocks of ammunition, the fighting, which was largely confined to Aden, was fierce and intense from the beginning.

The fighting was particularly heavy outside the Aden Frantel Hotel near the Tariq army camp, which was soon littered with the wreckage of armoured vehicles. Panic-stricken tank gunners fired in every direction as they tried to escape rockets launched from high buildings. Two ammunition dumps exploded near the Crater district.

As the street fighting continued, the Soviet, East German and Cuban advisers at least had the satisfaction of knowing that their Yemeni trainees had become proficient in their work. The navy, the larger part of which had chosen to fight for Ismail, was particularly adept in finding shore targets and hitting them from its 8 Soviet Osa II frigates while suffering little damage itself. Unable to stop the fighting, the Soviet commanders in South Yemen mounted strong guards on the Russian bases so that at least the battles involving heavy weapons would subside in time due to lack of ammunition. On the outbreak of the war 2,000 Soviet advisers and 10,000 military personnel as well as 400 Cubans and East Germans were in South Yemen but took no part in the fighting.

British public imagination was caught by the role of the royal yacht, *Britannia* (Rear-Admiral John Garnier). The ship was in the Red Sea on its way to New Zealand in preparation for a royal tour when the war broke out. With British and other civilians in danger, *Britannia* made several trips to a beach east of Aden to pick up evacuees. The ship's boats went in and collected the civilians waiting in the water. At one point a battle took place at the embarkation point and Admiral Garnier had to pull out his shore party. The evacuees, of several nationalities, were taken to Djibouti. According to the British Ambassador to South Yemen, Mr. Arthur Marshall, if *Britannia* had not gone to the rescue there would have been "a bloodbath of unthinkable proportions".

President Muhammad flew to Ethiopia to seek help from the Marxist regime of President Mengistu and was promised some arms, but it was doubtful if they could arrive in time and in safety. On 19 January Muhammad returned and landed at the airport of Taizz near North Yemen's southern border, from where he proceeded to his home region in Abyan, 100 miles east of Aden. By now his soldiers were fighting their last battles, including one in Crater, Aden.

By 23 January *Britannia* and several other ships had evacuated about 4,500 foreigners from various countries, including 3,000 Soviet citizens from the Aden area and Al Mukalla port, about 500 km to the east in Hadhramout governate.

By that time almost the entire city of Aden lay in ruins and 10,000 people, mostly civilians, had died. Total damage ran into several billion dollars.

The civil war in South Yemen was a political calamity for the Soviet Union and potentially a strategic disaster. It showed that the Russians were unable to control their closest ally in the Middle East and a country which is crucial to their diplomatic and military strategy. Vladimir Polyakov, head of the Near East department of the Soviet foreign ministry, wanted to use the Russian success in South Yemen to show other Arab states that co-operation with the Soviet could be profitable. South Yemen was supposed to be the model with which to woo all Arabs. The fighting also damaged the reputation of the KGB and the military intelligence agency; apparently both failed to foresee the conflict.

Rebuilding South Yemen's economy will be an enormous and vastly expensive task and much foreign money is necessary. The Soviet Union let it be known, even while the war was in progress, that it is in no financial position to offer the aid required. The Russians had already invested large amounts of money in South Yemen during President Muhammad's regime. The Soviet navy's Indian Ocean squadron, which includes nuclear-armed submarines, had secure operating facilities and naval infantry often practised amphibious landings on Socotra island. A flight of Soviet IL-38 anti-submarine planes had operated from Aden since 1983. The principal function of the South Yemen-based forces would be to cut the supply of oil from the Gulf to the West in time of war.

For North Yemen and the government of Ali Abdullah Saleh the situation was even more critical. It was difficult to see how North Yemen could avoid involvement in South Yemen's troubles. Sana'a studiously avoided any action that could be regarded as interference, but on 19 January the government allowed a group of Palestine Liberation fighters to travel to the border between the two countries. The PLO contingent was to act as a buffer force to support a ceasefire. South Yemen soldiers turned the PLO back. An additional complication for Sana'a was that some tribal territories extended across the frontier and tribespeople in the north were slipping across the border to support their fellow tribesmen in the south. A still further difficulty was that the overthrown president, Nasser Muhammad, had fled to North Yemen, there to build up an army to regain power in South Yemen.

The only consolation for the North Yemen government and for the Soviet Union was that Haydar Abu Bkr el Attas had been appointed President of South Yemen. A moderate, with close Kuwaiti and Saudi connections, el Attas had not tilted towards either YSP faction and from everybody's point of view seemed to be the ideal choice.

Interestingly, the new Yemeni leadership deliberately delayed announcing the death of Abdul Ismail—he was killed during the fighting—until a new government was formed. They feared a fresh outbreak of violence should the pro-Ismail mobs learn of his death. His funeral was attended by 15,000 mourners, including President Al Attas. Ismail was one of 13,000 victims of the power struggle.

In 1986 the civil war subsided into an Afghan-like power struggle among tribes. Traditionally, tribal warfare in both Yemens is difficult to contain. Sana, the capital of North Yemen, and Taizz, a border city, were rife with unrest following the fighting in Aden. It affected some southern provinces in Saudi Arabia, where the large tribe of Bani Qahtan had not settled down following the Saudi/North Yemen/South Yemen border fighting in 1979.

Whatever might happen in terms of tribal warfare, South Yemen will return to its hard-line Marxist policies.

Sri Lanka Civil War

TAMILS AGAINST SINHALESE

Background

The civil war in Sri Lanka between the Tamils and the Sinhalese, a classic conflict of insurgent guerrillas fighting government troops, became ferocious during 1985 and another minority group, the Muslims, became involved.

The ethnic animosity between the Buddhist Sinhalese and Hindu Tamils had increased steadily since Sri Lanka (formerly Ceylon) became independent from Britain in 1948. Demographically, the 11 million Sinhalese are dominant; the Tamils number perhaps 3 million and the Muslims 1.5 million. In a national area of 25,332 square miles the Tamils are heavily concentrated in only about 1,000 square miles, mainly in the north, but with smaller areas in the centre and on the east coast.

The Tamil demand for a separate state, which they would call Eelam, began in 1948 when Sinhala was proclaimed the official language. In addition to this advantage, the Sinhalese gained greater control over the universities, the police and the armies. In 1977 Tamil was recognised as the second official language, but this did not mollify the Tamil separatists, especially when Buddhism later became the official state religion.

When Sinhalese and Tamil civilians fought each other in 1983, the army deployed troops in the Tamil north "to protect civilians". A cycle of bombings and reprisals began. Young Tamils from the Jaffna district openly resisted what they termed the "occupation army". More and more acts of violence were committed against the police. The guerrillas claimed to be freedom fighters, and young Tamils oppose the "excessively moderate" stance of members of Parliament belonging to the Tamil United Liberation Front (TULF) led by A. Amirthalingam.

TULF is only one of at least 23 different Tamil groups. The best organised militarily is that of the Liberation Tigers of Tamil Eelam (LTTE). Other leading organisations are the People's Liberation Organisation of Tamil Eelam (PLOTE), the Tamil Eelam Liberation Organisation (TELO), the Eelam People's Revolutionary Liberation Front (EPRLF) and the Eelam Revolutionary Organisation and Supporters (EROS) founded in London in 1975 by Eliathamby Ratnasabapathy.

The members of the terrorist groups are known collectively by the people and the press as "the boys", and they number about 6,500 in all.

The boys want Eelam to be a completely separate Tamil state. The men of TULF will settle for much less, but they do insist that they want Tamil districts to control internally two key areas of power, the police force and land use. Both measures, says TULF, would make Tamils feel more secure. A Tamil police force would remove a major source of ethnic conflict and oppression. Tamil control of land would preserve the ethnic balance and prevent encroachment by the Sinhalese. They also want the Tamil districts to be administered as one province; this means

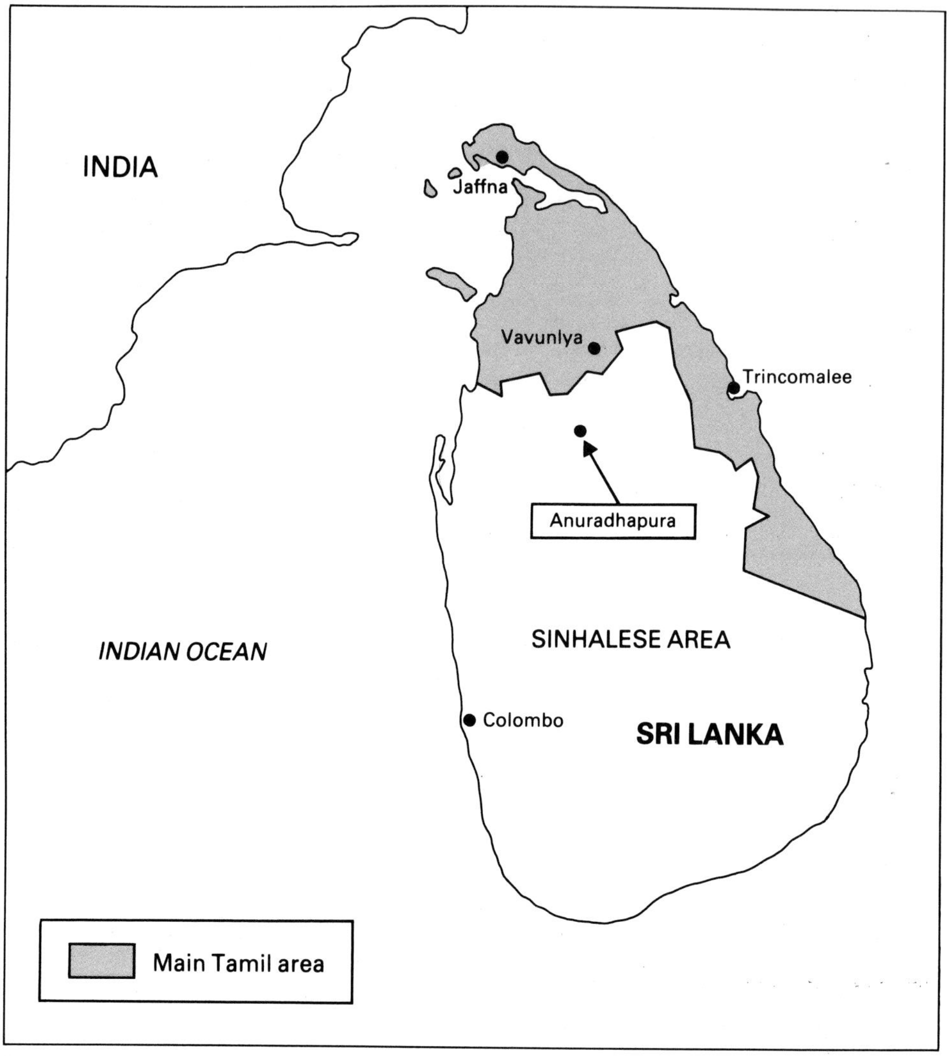

Sri Lanka Civil War

giving the Tamils an area equivalent to the notional Eelam—the northern and eastern provinces combined to form one unit with Trincomalee as its capital.

The government of President J. R. Jayawardene is implacably opposed to the idea of a Tamil state in Sri Lanka. For them a Tamil state already exists just across Palk Strait; it is the Indian state of Tamil Nadu, with Madras as its capital. The Sinhalese believe that creation of a Sri Lankan Tamil state would be quickly enlarged by a flood of Tamil immigrants from across the strait. In their fierce reaction to Tamil separatism, the Sinhalese killed thousands of people in the anti-Tamil pogroms of 1983.

Meanwhile, the many different Tamil groups set up their headquarters in Madras to fight what can only be described as state terrorism. On 4 August, at Valvettiturai, sailors from the Sri Lankan Navy were involved in a clash with fishermen; two sailors and a policeman were killed. A short time later, army tanks and jeeps moved into the town and ordered all men between the ages of 18 and 25 to come out with their identity cards or risk being killed. They were taken away and the town destroyed; the men have not been heard of since.

The following month, a young woman was killed in her sleep by a shell fired into a small village from a navy ship lying a few miles offshore. On 11 September a bus carrying Tamils was attacked by 6 soldiers; they slaughtered all 16 men on the bus. More than 2,000 people died during the year.

By the end of 1984 Jaffna was bandit country. No tourist ventured there, and consulates in Colombo had recalled administrative personnel and peace-corps workers. In December 1984 a landmine killed the chief of the army's northern command, Colonel Ariapperuma.

The War in 1985

Sri Lanka has no military tradition, mainly because the British as the colonial masters did not create standing armies in Ceylon as they did in mainland India. The units raised by the British tea planters did little more than play at soldiering. In addition, after 1948 the new Sri Lanka did not feel threatened and unlike many newly independent nations considered it unnecessary to build up its armed forces.

The civil war which gradually developed found both sides militarily unprepared. The government turned to the US for financial aid with which to buy arms and equipment and to Israel for training. Sri Lankan–Israeli relations had been broken in 1970, but in 1984 President Jayawardene's government asked for Israeli help to counter the Tamil insurgency. An Israeli Interests Section was established in the American Embassy in May 1984 and Israeli experts arrived to help train the police and advise on intelligence gathering.

A special task force of police commandos was created in 1984 and trained by British experts who are former members of the Special Air Services (SAS). The commando unit has been accused of appalling atrocities; as a routine punishment they burn the homes of families suspected of harbouring terrorists.

The Sri Lanka army is so ill equipped that in 1985 the government launched a "help the forces" campaign, urging members of the public to give generously to help the army, navy and air force to eradicate terrorism. This is not just a matter of supplying "comforts"; citizens have been asked for rain capes, water bottles, beds, buckets, crowbars, refrigerators, ladders, even wheelbarrows.

Until recently, the security forces had no armoured helicopters and their only ground-attack aircraft were a few Cessnas equipped with machine-guns. Their patrol boats are easily outpaced by fast fishing craft crewed by Tamils based in India.

"The boys" turned for help to their own Tamil community in southern India and to India generally. At least one radical group, PLOTE, sent its members to a camp near Damascus where they were trained by the Marxist Popular Front for the Liberation of Palestine (PFLP). The PFLP's ideology would not disturb the Tamil guerrillas, most of whose leaders are Marxists themselves.

By early 1985, the Tamil groups had evolved a three-part strategy for increasing unrest in Sri Lanka, with the object of bringing down the government.

1. Attacking the government forces by ambush and landmines and surprise assault.
2. Stealing and burning state property.
3. Attacking Sinhalese civilians whose property encroaches on what the terrorists believe to be traditional Tamil homelands.

Each Tamil group is made of three branches—a military unit, a propaganda cell and a pool of sympathisers. Also, each group seems to specialise in a particular type of attack and target, a pattern apparently suggested to the Tamils by their Palestinian friends. For instance, TELO specialises in spectacular coups; they blew up a Sinhalese troop train, killing 60 soldiers. EROS concentrates on economic sabotage. This group set off a bomb at Oberoi Hotel in Colombo, the capital, and another at the Ministry of National Security. PLOTE, the richest of the groups—its funds could be coming from Libya—is the least active in overt operations. Terrorist actions are pointless, PLOTE leaders say, because Tamil settlements cannot be protected from army reprisals. Therefore, PLOTE tactics are to harass the government and individual leaders politically as well as by threat and blackmail.

The 2,000 Tigers are fighters. This faction, which draws its title from an old Tamil legend, is the "image group" in the minds of the Tamil populace. Their leader, Prabhakaran, has trained the Tigers well and made them into a disciplined force. They have mortars and Sten guns which have been taken from "the enemy" or bought on the open European arms market. Their Kalashnikov AK47s came from the PLO. The funds are provided by Tamils living in Europe and the Middle East.

In May 1985 terrorists attacked the Sri Lankan naval base at Karainagar, 12 miles north-west of Jaffna, and killed several naval personnel. Simultaneously, another group attacked the army in the old Dutch fort of Jaffna. Tamil aggression has been as imaginative as it has been bloody. In May 1985 a busload of terrorists disguised as soldiers, and travelling in a bus, opened fire in Anuradhapura, a sacred pilgrim city in the North Central Province. They killed 146 civilians and injured hundreds more. Through loud-hailers the Tamils announced that their act of murder and mayhem was in retaliation for an army rampage against Tamils.

Government retaliation was swift. Attackers boarded a ferry off the northern coast of Sri Lanka and hacked 39 Tamils to death with axes, swords and knives. The Sri Lankan Navy denied that its men were responsible, but nobody believes the denial. Brutal dawn raids are commonplace.

Trained in weaponry at camps across Palk Straits in India, the boys mount daring raids in fast boats, they blow up barracks and mine roads. On one occasion they slaughtered about 80 Sinhalese who had settled on what the boys considered to be Tamil land.

Tamil tactics include raids on foreign development projects, a move designed to bring foreign pressure to bear on the Sri Lankan government. In one exploit the Tamils attacked the giant Mahaweli irrigation scheme, a US aid project for which Britain provided an outright grant of £100 million.

A Tamil raiding party was confident enough to attack the Vavunlya cantonment to free 60 Tamil guerrillas held prisoner there. They inflicted casualties on the guards and escaped into the jungle with the former captives and without casualties of their own.

Studying at Jaffna University, one of the best universities in Asia, has become difficult. Some active supporters of Eelam have dropped out and are living secretly in their villages or are trying to flee to India. Others are in touch with the Tamil communities in the centre of the island and trying to win them over to the cause. They are taking advantage of the extreme poverty and the disastrous situation of the Tamils working in the tea plantations; the Tigers are trying to recruit them and win the sympathies of the younger generation.

The morale of the Tamils is higher than that of the government troops, many of whom are young, ill-disciplined and terrified. At least 200 soldiers have been accidentally killed by their comrades firing wildly in the dark at supposed enemy.

But while the Tamil fighters have to be opportunist in their tactics, the government forces are more deliberate. For instance, the army decided to make the island of Mannar, where several Tamil groups had training camps, into a no-go area. The troops used napalm to burn down the groves of coconut palms and palmyrah while other forests were cut down by gangs with saws. Mannar was turned into a huge sand dune, so that it cannot provide cover for "the boys", about 200 of whom were killed in an army sweep.

The army's raids are devastatingly destructive; a prosperous Tamil village can be turned into smoking wreckage within half an hour. The security forces' co-ordinating officer for the Trincomalee region, Commodore Jayasuria, is particularly efficient. He commands 10 battalions of infantry, commando squads and naval assault parties as well as helicopters and reconnaissance aircraft. In one series of lightning raids Jayasuria's troops attacked 23 villages and set light to 1,000 homes; more than 100 people were killed in these attacks. Each night Tamils are shot when caught in the street after curfew.

The struggle for Trincomalee is critical. For the Tamils, it is the natural capital city of any independent Tamil state. It is important to the Sri Lanka government as a naval base and as an expanding centre for container trade with the Orient.

During the last year the Indian authorities appear to have turned against Tamil terrorism by confiscating weapon consignments. On one occasion a small boat was carrying cases of British-made grenades, machine-guns and rifles from the tiny Tamil village of Mandapam on India's Tamil coast to the "front" on the Jaffna peninsula. An Indian coastguard boat picked out the small boat by searchlight, seized the arms and arrested the crew. A week later a container as big as a railway wagon arrived in Madras from Singapore. It was marked "waste paper", but Indian Customs found in it 1,400 automatic rifles, 300 Sten guns and Japanese field radios. The arms were intended for PLOTE.

These unfriendly acts alarmed the Tamil leaders and they sent emissaries to the Indian government to ask for co-operation, even invasion, for the Tamil separatist movement. They were told that there would be no invasion. In fact, India had earlier been discreetly told that such a move would result in a condemnation by the UN Security Council.

Anti-Tamil government propaganda has brought Sri Lanka's Muslim people into the conflict. Most of the 1.5 million Muslims live in the eastern province, where they make up 30% of the population. They were stirred into violence by Sinhalese *agents provacateur* who told them that the Tamils want to outlaw the Islamic religion and to seize Muslim land. Mobs of Muslims have attacked Tamil shops and houses. At Karaitivu, thousands of Tamils were made homeless and an unknown number killed in a Muslim attack on homes and a Hindu temple.

No matter how strong the armed forces become, and how much anti-Tamil feeling the government can generate among Sinhalese and Muslims, "the boys" are unlikely to be defeated or forced to retreat to India. They are intelligent, determined and ruthless. At the end of 1985 the freedom fighters (or terrorists) had become true guerrillas; that is, they were operating without fixed bases—other than those in India. The armed forces cannot find them and can only punish their sympathisers. And because of the army's atrocities, the guerrillas' violent tactics are widely accepted by the Tamil populace.

State Terrorism

In 1986 the army sent a special task force into eastern Sri Lanka to use sustained terror in an effort to break the Tamils' morale. The town and district of Batticaloa were chosen as the "target area." The specific targets were Tamil youths, many of whom have been killed or have disappeared. A Catholic priest who works with the Batticaloa citizens committee told the few foreign reporters able to get through to the area, "The situation is so bad that all Tamil males in the 14–17 age group must leave the eastern province if a slow genocide is to be prevented."

A typical incident happened in the Iruthyupuram suburb of Batticaloa. Before dawn a hand grenade was thrown at a patrol and two policemen were killed; the Tamil militants responsible then committed suicide by taking cyanide tablets. The security forces dragged hundreds of Tamil males from their homes, herded them into a Catholic church where they were savagely beaten up; 24 young Tamils were killed.

Every incident of this kind leads to a reprisal by Tamil fighters, such as the attack in Dehiwatte near Trincomalee in which 39 people, nearly all Sinhalese civilians, were killed when their lorry went over a landmine planted by Tamil terrorists.

During the first half of 1986 government planes air-bombed the Jaffna district almost every day. This was part of the government's campaign to achieve, as President Jayewardene put it, "a military solution" to the Tamil insurrection. More than 97 per cent of the population of Jaffna is Tamil. Therefore, by the government's reasoning, anybody killed by air attack was certain to be a terrorist. In fact, the majority of Tamils just want to be left in peace. The government had ignored an old lesson of guerrilla warfare that victory is seldom attainable by air power alone.

Sudan Civil War

VAST BATTLEFIELD

Background

An earlier war was fought between the years 1955 and 1972 between the Muslim Arabs of the north and the Christian negroes of the south, following Sudanese independence from British rule. The worst fighting occurred between November 1958 and October 1964 during the military regime of General Ibrahim Abboudd. In 1963 the southerners, who had been brutally oppressed by the Arab army from the north, formed the Anyanya ("venom of the viper") guerrilla army. When General Gaafer Nimeiri came to power in a coup in 1971, the Soviet Union gave him military aid and Russian bombers flown by Russian pilots bombed Anyanya strongholds. Following efforts by the World Council of Churches and Emperor Haile Selassie of Ethiopia, the war formally ended in March 1972.

However, Islamic pressure and army oppression from the north forced the Christians of the south to continue their guerrilla activity. They built up their own semi-guerrilla army, the Sudan People's Liberation Army, whose leader in 1982 was Colonel John Garang, aged 42, with a PhD from Iowa State University.

He found allies in Ethiopia and Libya, both of whom wanted to undermine the Nimeiri government. Ethiopia, on Soviet instructions, wanted to bring Sudan into the Soviet camp; Colonel Gaddafi of Libya would like to add Sudan to his Islamic empire. Gaddafi also wants to break Sudan's ties with Egypt, one of his main enemies.

That Gaddafi, an Islamic fanatic, should help Garang, a Christian, might seem strange; however, Gaddafi planned to deal with Garang once he took over Sudan with Libyan troops.

In the meantime, a second Anyanya organisation, known as Anyanya II, led by William Abdullah Chol, had become a rival force to John Garang's SPLA. It seemed more amenable to co-operation with the government, whereas SPLA wanted to secede.

Garang refused all approaches by intermediaries trying to achieve a negotiated peace between him and Nimeiri. The British Foreign Office, the American State Department and Chevron, the US company which was developing Sudan's oil—which is all in the south—made numerous vain approaches. The Kenyan government, British businessman "Tiny" Rowland and the Arab oil magnate Adnan Khashoggi also joined in the peace efforts.

The War in 1985

During 1984 and 1985 Garang revealed great skill as a guerrilla leader. Apart from his hit-and-run tactics against the Sudanese Army, he showed his understand-

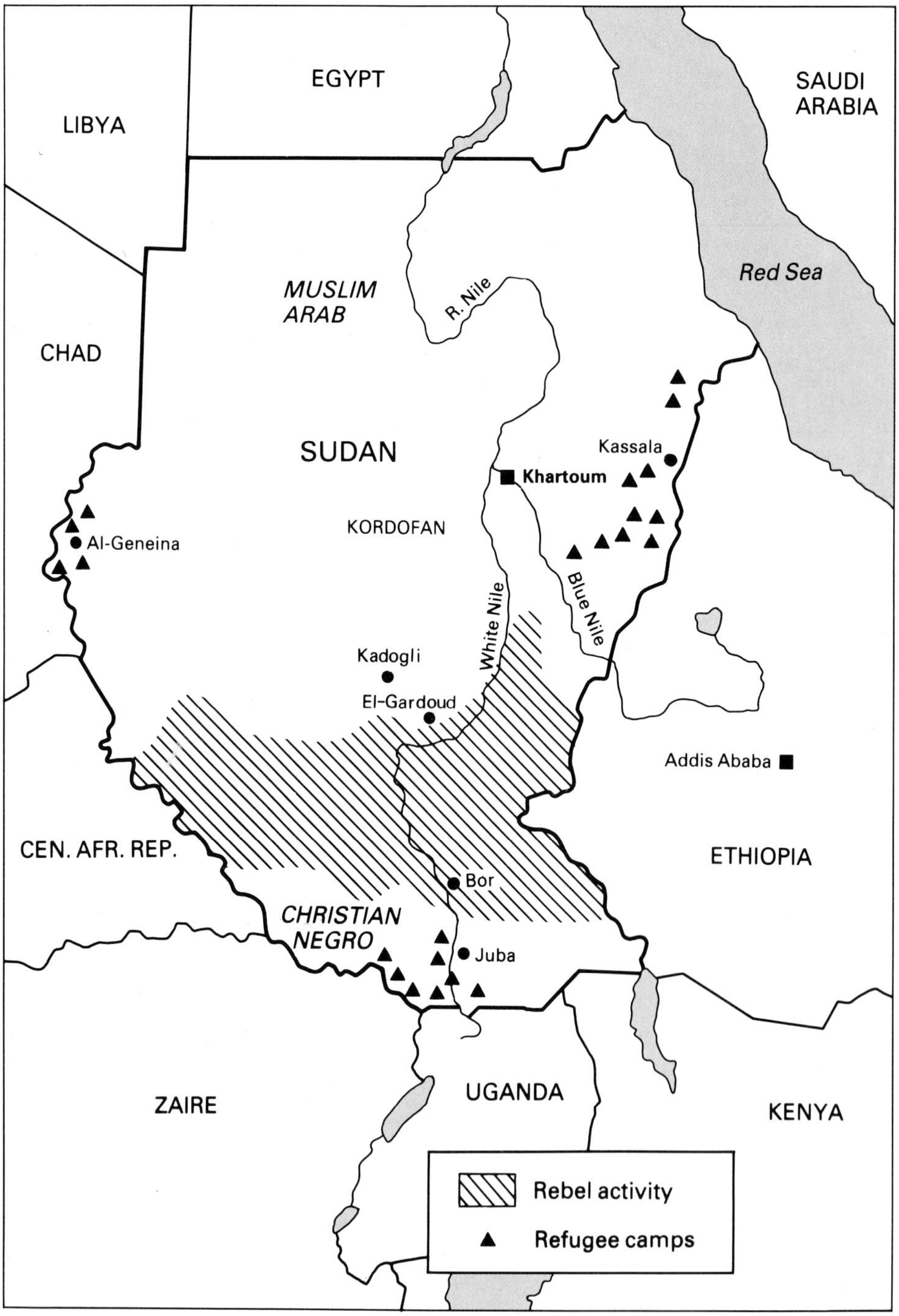

Sudan Civil War

ing of the effects of damaging the national infrastructure. By sabotage, kidnapping and threats he closed down work on the oilfields and on the great Jonglei Canal project which would vastly increase Sudan's agricultural yield.

Kenya's president, Daniel Arab Moi, intervened as a negotiator. He was prepared to guarantee both Garang's safety if he travelled to Khartoum to talk peace terms and the political terms of any agreement reached. Kenya would gain from a Sudanese settlement, for when oil starts to flow and the Jonglei Canal is completed, pipeline and rail links though Kenya could boost the importance and wealth of the port of Mombasa. Garang rejected Moi's overtures.

His proclaimed aim is to take over power in the whole of the Sudan, but he knows that as a Christian he had no realistic chance of doing this; Muslims outnumber Christians four to one. More practically, he can become "Tzar of the South"—as he is already called—or president of a new Christian state.

With fewer than 10,000 guerrilla troops, he controlled by early 1985 two of southern Sudan's three provinces. Malakal, capital of Upper Nile province, was surrounded by his men and he was in a position to shoot down supply aircraft flying into Juba, capital of Equatoria province. In addition, SPLA's powerful radio station broadcasts propaganda, in Arabic and English, into Khartoum every afternoon.

As a demonstration of his military prowess and power, Garang ambushed and sank two Nile steamers; in this attack his men killed 150 people and captured 300 Sudanese, Tanzanians, Ugandans, Kenyans and Zaireans. In February 1985 SPLA overran Jeeko, an important point between Nasir and Madan on the Upper Nile Province, and raided Malwal, the army defence centre for an oil-drilling region. The SPLA units was able to cut roads and disorganise the army at will.

On 5 April 1985 President Nimeiri was overthrown in a military coup and was replaced by a "transitional military council" (TMC) led by General Swar al-Dhahab. The new president and his colleagues announced that they were having talks with Garang and his SPLA, but Garang denied any negotiations. Indeed, for the first time since 1983 SPLA extended its operations into northern Sudan. In late May it attacked El-Gardoud, 75 miles south-east of Kadogli in southern Kordofan. The commander of the nearby garrison on Talodi rushed to the rescue and himself became a casualty. The SPLA drove off 4,000 cattle and many sheep, in keeping with its consistent policy of living off the land—and where possible enemy land. In June SPLA gunners sank two river steamers and shelled the provincial town of Bor in southern Sudan, killing the garrison commander.

The army command has never revealed that SPLA guerrillas have raided air force bases near Khartoum, thousands of miles from their own territory. In one audacious raid by road they destroyed 4 Mi-21 aircraft. On another occasion they damaged three aircraft, probably C-130Hs, of the air force's only transport squadron. At a time when very few army officers knew that the government had ordered 6 Bell 212 helicopters, SPLA's radio announced that they would be put out of action if they ventured into southern Sudan. It has also not been revealed that the Sudan Air Force has never made a positive sighting of SPLA guerrillas, who can count on the support of all the southern Christians and the animist tribes who live in the south. Towns and army posts under SPLA attack generally call for air support, but by the time the aircraft arrive the guerrillas have long gone.

The list of casualties in the southern fighting lengthened throughout 1985. An officer of the commando unit based in Jebel Aulia, outside Khartoum, revealed that the army rarely informs relatives of deaths, making it impossible for them to make any financial claims. This demoralises the army.

The TMC refused to reinstate officers who claim that they were unfairly dismissed by Nimeiri. They cut the size of the army from 53,000 to 50,000. The deteriorating military situation so worried some senior officers that in June a group of senior men attempted to overthrow the Swar al-Dhahab regime. Most of the 11 officers concerned in the abortive coup belonged to the Arab Baath Socialist Party and a few were Communists.

The TMC is unwilling to allow the south to secede; in an area the size of France, it contains Sudan's most promising oilfields. That source of revenue is badly needed by a country that has been transformed by gross mismanagement and corruption from a potential exporter of food into one which desperately needs international assistance.

Anyanya II's leader, William Chol, had agreed to seek a separate negotiated settlement. Soon after the overthrow of Nimeiri, he sent a high-level delegation to Khartoum to affirm Anyanya II's commitment to a such a peace, and with his men he fought alongside government forces. He was killed in August during a battle with SPLA troops in Fangak, Upper Nile region. This seriously damaged the government's hopes of peace, especially when many of Chol's followers joined Garang's SPLA.

There is no obvious end to the war. Sudan's armed forces are not strong enough to defeat SPLA, even if Garang's men could be found. In Sudan's 1 million square miles, the Sudan Air Force's 34 combat aircraft and its single helicopter squadron (9 Puma, 10 BO-105 and 5 Bell 212) is quite inadequate for the task of search and destroy.

The SPLA leaders showed their confidence in 1986 when they changed from guerrilla to conventional methods of attack. They shelled and then, in large numbers, assaulted army garrisons. On 5 March they captured the garrison town of Rumbek, a significant achievement. According to a report by Major General Muhammad Zein Al Abdin, army commander in the south, the rebel SPLA not infrequently used up to 3,000 men to overwhelm an army company of 200. Garang's intention was to demoralise the army by instilling in the soldiers a sheer fear of annihilation. Garang also gave the SPLA a more conventional structure by appointing a chief-of-staff. He wanted to convince middle-class Sudanese that his army was a more responsible one than the national army.

Responding to guerrilla attacks, General Abdin used fewer mechanised columns and went back to foot patrols to reduce casualties from SPLA ambushes.

During 1986 the propaganda war became more intense and more sophisticated. Khartoum portrayed the SPLA as dominated by the cattle-herding Dinka tribe, to discourage recruits from other southern tribes which share the same grievances against the Arab north but fear excessive Dinka influence. The SPLA projected itself as a national movement. Church leaders in the south spoke well of the SPLA's discipline and leadership.

Sudan held limited democratic elections in April 1986. The country's foremost political figure, Sadiq el-Mahdi, backed the government and the army and argued that Garang was "completely subordinated" to Marxist Ethiopia. Despite Sadiq's support it was unlikely that General Swar al-Dhabab would fully hand over to a civilian government while a civil war continued to divide the country and with the ever present danger that the deposed Nimeiri might stage a counter-coup.

Thailand Attempted Coup

For Colonel Monoon Roopkackorn of the Thai Army and his more senior supporters, 9 September 1985 would have seemed like the perfect timing for a coup. The king was on holiday in his southern palace near the Malay border; the prime minister, Prem Tinsulond, was visiting Indonesia, and the supreme military commander, General Arthit Kamlangek, was in Sweden. The crown prince, a lieutenant-colonel in the palace guards, was on his way to Italy.

Colonel Roopkackorn, a former commander of the 4th Tank Regiment, had been dismissed from the army for organising an earlier unsuccessful coup against General Prem Tinsulond on 1 April 1981. This time he and his supporters commandeered 22 US-made M41 tanks which they placed in front of the supreme command HQ. Roopkackorn's brother, a wing commander, brought in some air force security troops so that the colonel had in all about 500 men.

They occupied HQ and the civilian radio station and called for the army to rally behind General Serm na Nakhon, a retired supreme commander and former deputy prime minister. No support materialised. Pro-government generals, over army radio, ordered the rebels to surrender and prepared a counter-attack with the crack 1st Division.

When the rebels shelled the army radio station and then exchanged small-arms fire with its guards, 5 people were killed and 120 wounded, most of them civilians; they included two journalists from American NBC television. The rebellion was over in 10 hours and Colonel Roopkackorn was flown to Singapore from where he asked for asylum in the US. His wing commander brother "disappeared".

Two other senior officers were implicated—General Kriangsak Chomonan, a former prime minister, and General Yos Thephsadin.

Since 1932, Thailand has had 8 successful coups and 8 failed coups. The popularity of King Bhumibol Adulyadej, who has been on the throne for 39 years, was perhaps the main reason for the 1985 coup's failure. The civilian-military government of Mr. Prem Tinsulond (he had to resign from the army to enter politics) has been in power since 1980 and is the longest serving and least repressive in Thailand's history.

The Thai government was quick to assure its allies, particularly fellow members of ASEAN and the US, that the events were nothing more than a throwback to a more troubled period.

The generals are more concerned with the Vietnam-Kampuchea-China conflict than with internal squabbles. The army, 235,000 strong, has had some major fights with Vietnamese army units attacking Kampuchean resistance groups which have bases inside Thailand.

Neither the Thai political nor military leaders want a war with Vietnam; it would

Soldiers loyal to Thailand's government approach the headquarters of the coup leaders on 9 September 1985; the rebellion was crushed soon after.

be bloody, destructive and Thailand would probably lose. However, there is a limit to Thai patience. Late in 1985 one of the army's most important leaders, General Chaovalit, said that Thailand will take "whatever action is necessary" to stop attacks on Thailand, including pre-emptive strikes against Vietnamese units in Kampuchea.

Uganda Guerrilla War

A NATION DESTROYS ITSELF

Background

In 1980 President Milton Obote swept to power for the second time after the dreadful years of Idi Amin's cruel and destructive regime. The former defence minister, Yoweri Museveni, claimed that the election had been rigged—which seems very likely—and with 26 supporters took to the bush to wage a guerrilla war against Obote.

Britain and the Commonwealth sent military training teams to Uganda and gave a lead to the World Bank and other donors who set about rehabilitating the economy under Obote. According to several critics, London hushed up Obote's human rights violations and bears a heavy responsibility for the massacres during his regime.

Museveni, meanwhile, styling his force the National Resistance Army, gathered 6,000 fighters, many from the largest and best educated tribe, the Bagandans. By 1982 the NRA, which was largely armed by Libya, controlled the Luwero Triangle to the north and west of Kampala, a highly populated area producing most of Uganda's coffee crop. The government sent the Ugandan Army into the Triangle where the soldiers were guilty of many atrocities against people they suspected of having helped the NRA. In 1984 Elliot Abrams, Assistant US Secretary of State, said that more than 200,000 people had been killed in the area by government forces. The real figure is more like 300,000.

The War in 1985

Army atrocities in the Triangle continued. Villages became empty, buildings crumbled and the coffee plantations were overgrown. The few remaining civilians were loyal to the guerrillas. The Obote regime claimed that the army was under control and that there were fewer atrocities; in fact, there were few people left to kill.

Museveni's tactics with his steadily strengthening NRA were to move out of his area to ambush and attack the Ugandan military, with occasional forays into Kampala. Assassination attempts and bomb attacks on government targets became common.

Late in June 1985 the army chief-of-staff, General Smith Open Acak, loyal to Obote and his Langi tribe, ordered a group of Acholi soldiers to be sent to the front line to counter the growing threat from the NRA around the town of Fort Portal. In the large Magamaga barracks, where the 13 British instructors were stationed at Jinga, east of Kampala, the order started a shoot-out between the Acholi and Langi. These two northern ethnic groups have dominated the army since colonial times and were the basis of Obote's power.

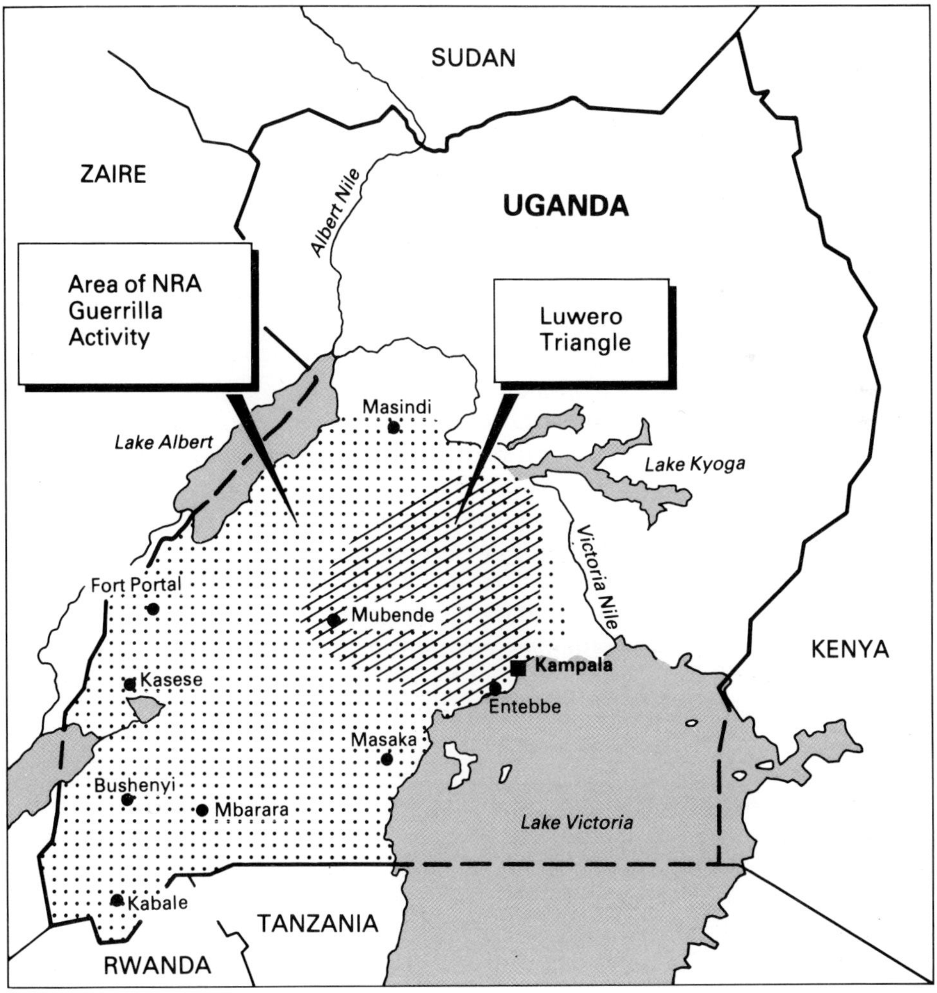

Uganda Guerrilla War

On 27 July Brigadier Basilio Olara Okello overthrew Obote in a coup. He then asked General Tito Okello, a retired officer aged 71, to become president. The two men are not related, but are both members of the Acholi tribe. It was the belief that Obote discriminated against the Acholis and favoured officers of his own Lango tribe that helped to cause the coup.

Confusion and fear shook Kampala after the Okello takeover. Bands of drunken soldiers armed with Kalashnikovs roamed through the city, looting stores, stealing cars and harassing Ugandans and foreigners alike. About 60 people were killed and several British residents were beaten.

Okello, quickly installed as president, imposed a curfew and appointed a prime minister, Paulo Muwanga, former vice-president and minister of defence under Obote. This caused suspicion and alarm among many Ugandans.

Museveni and his NRA, which by now had grown to 8,000, did not immediately side with the new regime, as it had expected. But Museveni took advantage of the situation to move out of the Triangle. He consolidated his hold on Fort Portal and extended his military operations through a swathe of territory that covered more than a third of Uganda's area. Within 6 weeks the guerrillas controlled Masaka, Kasese, Mbarara, Mushenyi, Kabale and Masindi.

In some of these places soldiers of the Ugandan Army linked with the NRA or did not put up much resistance before being disarmed. Museveni announced a ceasefire, but sent a message to Okello that it would hold only if his conditions were met. The NRA issued a statement that it intended to restore democracy and security to Uganda. It had shown for 4 years that it was not afraid of the army, and Museveni claimed that the NRA had reached a stage of development where it could employ brigade and battalion strength units.

The NRA troops had certainly shown that they were capable of waging mobile warfare—as distinct from hit-and-run guerrilla warfare—and Museveni was prepared to fight what he called "positional warfare".

His guerrilla army had a variety of weapons, including light and medium machine-guns, a few heavy machine-guns and some artillery pieces; most had been captured from the Ugandan Army. Personal weapons include Kalashnikovs, Uzis, 9-mm model 12 Beretta sub-machine-guns, automatic pistols and grenades. Gaddafi stopped the Libyan equipment supply in 1982 when Museveni would not agree to the conditions Gaddafi wanted to impose; for instance, he insisted that Museveni have a Libyan chief-of-staff and Intelligence officers.

The NRA received few supplies from outside Uganda, and it was difficult for foreign arms to reach the areas Museveni controlled. Following the July coup he moved closer to the Zaire border, but neither Zaire nor Kenya wanted any involvement in Uganda's problems.

Despite Museveni's claim that he could fight a pitched battle, he did not have the necessary reserves of ammunition, but after the July coup officers who deserted from the Ugandan Army brought some supplies with them.

The NRA officers were all well trained either by the Tanzanians, whose army drove Amin from Uganda, or Frelimo, the Mozambique guerrilla organisation. Museveni was himself Frelimo-trained. These leaders had passed on this training to NRA recruits, who appeared to be better disciplined, better trained and more strongly motivated than the soldiers of the regular army, despite the work of the 13 British and 300 North Korean instructors.

Within 6 weeks of Obote's downfall, pro-Amin bush fighters from the West Nile region began to come freely into Kampala. Their leader, General Lumago, handed his weapons over to the Ugandan Army at a public ceremony, but he had access to other weapons. About 5,000 experienced guerrillas have been waging a war on Amin's behalf from the Zaire and Sudan borders since Amin was thrown out in 1979. They keep alive the fear among other Ugandans that Amin could return to power.

Without Museveni's support, the Okello-Muwanga government could not last. The rebels demanded a greater say in the formation of a government and strongly objected to Muwanga's appointment and to that of Mr. Wilson Toko as defence minister. He was Amin's air force commander until 1975 and had a bad name.

The Ugandan war again showed that a well-led guerrilla army, operating in bush and mountainous country and with the passive support of the civil population, could resist indefinitely against regular forces. Even virtual extermination of the civilians—which was Obote's policy—cannot weaken the guerrilla army under able leaders.

Another Campaign, Another Leader

General Okello and his supporters made the mistake of treating the NRA as just another armed group and failed to take into account the decisive role it had played in the collapse of the previous regime. Even more damaging for the Okello group's credibility was its obvious inability to have its orders obeyed by the regular soldiers. They went on committing acts of violence against the civilian population, in contrast to the tight discipline within the NRA's ranks. Musaveni had trained his men to respect the lives and property of villagers in the Luwero Triangle.

The NRA forces easily captured Kampala on 26 January 1986 and Museveni was sworn in as President three days later. He at once began work on forming a new national army and said that "rogues" would not be allowed to join. Potential recruits would be screened by village committees similar to those that operated in the Luwero Triangle during the bush war as security and intelligence bodies.

Museveni, a Munyankole from the south-west, is the first southerner to hold presidential power. Most of his officers come from tribes in the west of the country but the largest number of recruits are Bugandans, the most populous tribal group which forms the bulk of the population in the Luwero Triangle.

There can be no guarantee of continued peace in Uganda. Within weeks of Museveni's victory dissident northern troops started a guerrilla campaign of the kind which brought Museveni to power. Young boys of the 800,000-strong warrior Acholi tribe were being armed to fight, in imitation of Museveni's renowned policy of training eight-year-olds. Undisputed leader of the northern resistance was General Basilio Okello, based at Gulu, the regional capital. He might copy Museveni's strategy, tactics and training but he is unlikely to be able to command the same degree of respect.

War in Western Sahara

CONFLICT IN THE DESERT

Background

Western Sahara had been a Spanish colony from 1884. In December 1965 the UN called on Spain to decolonise its Spanish Sahara possessions and the following year urged the government to hold a referendum to permit the Western Sahara people to determine their own future. After further international pressure, Spain told the UN in August 1974 that a referendum would take place in 1975. This was postponed to permit the International Court of Justice to give a judgement on claims to the territory by Morocco and Mauretania.

The Mauretanians cited historical links and human settlement. The Moroccans based their claim on religious, juridical, demographic and cultural factors. The attractions of the region for outsiders are the high-quality phosphates at Bu Craa, the deposits of iron ore at La'youn and the considerable fishing prospects offshore.

The International Court, in October 1975, found that neither country had a legitimate claim and that the over-riding principle at stake was that of self-determination by the 75,000 Western Saharans.

The following month 350,000 unarmed Moroccans took part in the "Green March" into the Sahara to demonstrate popular feeling that Morocco should have the territory. In accordance with UN resolutions—but after having warned the world of the consequences—Spain evacuated the region in February 1976.

Morocco and Mauretania at once partitioned the territory, while the Western Saharan organisation, Popular Front for the Liberation of Saguia el-Hamra and Rio de Oro (Polisario), declared the whole area to be the Sahrawi Arab Democratic Republic (SADR). It was recognised by many countries, including a majority of members of the Organisation of African Unity (OAU), and began to wage a guerrilla war against the occupying Moroccan and Mauretanian troops.

In July 1978, after several military reverses, Mauretania agreed to a ceasefire, and a peace treaty the following year, renouncing all claims in Western Sahara. Morocco declared the entire area to be a Moroccan province, and induced certain tribal leaders to swear allegiance to King Hassan.

After 1979 numerous battles took place between Polisario and Moroccan forces. Polisario, which had only 3,500 fighting men, began as a highly mobile guerrilla force, preferring hit-and-run tactics, but it has also used conventional methods of war. Moroccan strategy became progressively geared to the "Hassan Wall", which was begun in September 1980.

Under Moroccan pressure, the Western Saharan people became concentrated either in the north-western area under Moroccan control or in refugee camps in neighbouring Algeria.

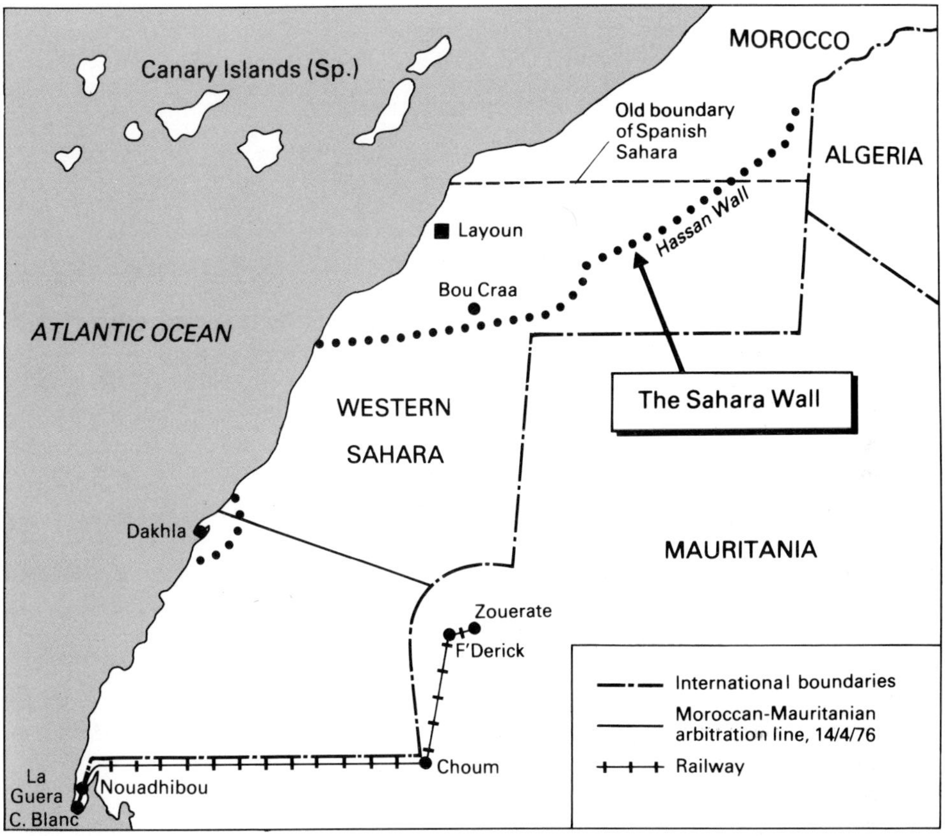

War in Western Sahara

At many conferences the OAU vainly attempted to negotiate a ceasefire and a settlement of the dispute. Morocco blocked the SADR's application for membership until February 1982 when an OAU Council of Ministers, in Addis Ababa, admitted SADR by majority vote. In August that year Morocco and its 18 supporters walked out of the annual OAU meeting, leaving it without a quorum. Subsequent attempts to bring about a settlement have foundered on Morocco's refusal to negotiate directly with Polisario.

The War in 1985

The Moroccan high command based its strategy on development of the Hassan Wall, a string of fortifications interspersed with minefields and designed to protect settlements and the more economically valuable parts of the Sahara. This is often called the "useful triangle" or "useful Sahara". This strategy limited Polisario's freedom of movement, and its fighters have never broken through the defensive position, though they have on occasions penetrated it.

The entire defence area covers an area of 35,000 km^2. The Hassan Wall, on which it is based, is 1,600 km long and is one of the most remarkable innovations in modern war. A wall of sand 3 m high and protected by barbed wire, the Hassan Line is, in effect, a tripwire. Its radars and ground sensors can detect an approaching enemy up to 30 miles away. Behind the wall, the quick-response units are poised to strike at intruders. The raiders are therefore robbed of surprise. The wall was planned with American help and equipped with American surveillance equipment; in return, Morocco gave the US permission to build bases in Morocco and to use military airport facilities.

Morocco propaganda claims with increasing vehemence that the Polisario fighters are no more than mercenaries in the pay of Algeria, which supports them politically and financially and on whose territory they find shelter. The Polisario fighters reject the mercenary sneer and stress with equal vehemence that their war is one of liberation.

Polisario has achieved no significant military success since 1981, and in 1985 it was clear that the war was going badly for the organisation. However, the guerrillas' traditional harassing tactics continued unabated with almost daily attacks on Moroccan positions. The Moroccan authorities deny that Polisario is more than a "flea-bite nuisance", but the truth is that the desert fighters, who are highly skilled and motivated, kill at least one Moroccan soldier each day; to achieve regular casualties against the enemy is a fundamental part of Polisario tactics. In this way the leaders hope to undermine Moroccan military morale.

SADR leaders are playing for time. They believe that Islamic fundamentalism will bring down the Moroccan monarchy and that the impoverished urban masses will become impossible for the authorities to control. Even the middle class, normally loyal to the king, is angry over the great cost of the war; it takes up 26% of the national budget. Soldiers on Polisario frontier duty receive double pay. In addition, the SADR is counting on continued Algerian support, though it lost Libyan support in 1985 when Morocco and Libya signed a "treaty of union".

It is possible, though, that time has run out for Polisario. During 1985 the Moroccan Army completed its take-over of virtually the entire northern territory, forcing the guerrillas into narrow zones hard up against the Algerian and

Mauretanian frontiers. The only place where the guerrillas can still roam at will is the deep south, on the stony plains of Rio de Oro. The desert here is so harsh and lifeless that the Moroccans have no immediate plans to occupy it.

SADR and its Polisario fighters say that they have the means, men and time to continue the war. Without the influx of French and American materiel, they claim, King Hassan would have given up his claims to Western Sahara long ago. Polisario has one powerful friend—the Soviet Union. During 1985 it supplied BMP-1 armoured vehicles. Armed with a 73 mm cannon, Sagger guided missiles, rockets and machine-guns, the air-conditioned and amphibious BMP-1 can carry a crew of 3 and 8 infantrymen; they fire through slits on the side, using individual periscopes.

In one action Polisario deployed about 50 Soviet-made T-55 tanks, 50 BMPs and perhaps 1,000 men—an immense concentration by the standards of guerrilla warfare. In a 5-hour fight at Haouza, the Polisario fighters used 122-mm artillery, 160-mm and 240-mm mortars, all Soviet-made.

However, Polisario's armour and other equipment is paltry when compared with Morocco's strength. Morocco has an army of 125,000 and paramilitary forces of 30,000, with several strong units trained to fight desert raiders. They include 5 camel corps battalions, 2 desert cavalry battalions, 3 commando battalions and 4 armoured car squadrons using AMX-10RC armoured cars. Late in 1985 the army took delivery of 126 VAB armoured personnel carriers, which are also intended for use in harrying the Polisario fighters. SADR leaders say that the Moroccan Air Force uses American-made cluster bombs against civilian targets.

Polisario, rather in desperation, has been sending "fighting patrols"—actually sabotage teams—into Morocco by sea, thus bypassing the Hassan Wall. They have caused damage to power stations and army depots, but these actions have no effect on the outcome of the war.

Campaign in 1986

SADR President, Muhammad Abdul Aziz, and his military chiefs made clear that Polisario would continue the war of attrition. Polisario's main tactic throughout 1986, they said, would be to breach the Hassan Wall and attack Moroccan command and artillery positions up to 6 miles behind it. Polisario can certainly achieve this aim but the damage to the Moroccans is more psychological than physical.

Even so, Morocco cannot easily destroy Polisario camps. The main one, at Tindouf, is protected by a wide array of Algerian-supplied anti-aircraft guns and missiles. There is also a strong garrison of Algerian troops and several squadrons of MiG fighters.

Even though Morocco is in a vastly superior position in its claims to the territory and in its ability to hold onto it, on the diplomatic front the two sides are a fair match.

War Trends

Helicopters and the Battlefield

A revolutionary change in tactical doctrine is taking place in the Soviet Union and the US because of helicopter technology. In the Soviet Union it is happening smoothly because the various services are not in conflict about roles and missions. Helicopters are part of Soviet Frontal Aviation (FA), and the ground forces collaborate closely, in line with the command structure. Administratively, FA is part of Soviet Air Force Command, but the ground forces control it operationally.

In the US, acceptance of the helicopter by all services in a close air-support role is still being debated. The army certainly accepts the helicopter, together with certain other aircraft, as an integral part of its combined arms team; the other parts are infantry, armour and field artillery.

The Russians refer to their rotary wing aircraft as combat helicopters and see them as indispensable on the battlefield. Extremely low flying, they appear unexpectedly to shower the enemy with bombs. In Afghanistan, the terrain does not lend itself to contour flying, but helicopters have still been the most effective way of fighting the hill guerrillas. Because of their experience in Afghanistan and their observation of battle in Angola, Ethiopa and Iraq, some Soviet planners believe that using supersonic aircraft for direct air support is a wasteful practice. Rather, combat helicopters, with artillery, have the task of hitting the enemy on the forward edge of battle.

In Soviet thinking, helicopters have a primary role in anti-armour operations; no major Soviet military exercise has been without them. Soviet doctrine, built on their own Afghanistan experience and that of Israel and Syria, is to use helicopters to attack tanks, armoured personnel carriers, self-propelled guns, bridge-layers and fuel trucks.

In any future conflict the Soviet armed forces will use helicopters to blow up bridges and lay mines. Newer helicopters will be fitted with missiles with automatic-guidance systems. The Russians are also producing a helicopter which they believe will be able to attack vertical-take-off-and-landing aircraft on the ground. Perhaps most importantly, the Soviet Army will use helicopters in the preliminary stages of combat to land troops to seize vital points and to confuse the enemy. Soviet tactical writings stress that it is better to have a company at a crucial point before the enemy can react than to have a division arriving afterwards. In accordance with this doctrine, elements of the 105th Soviet Guards Airborne Division seized key points in Kabul at the start of the Russian invasion in 1979.

The fact that the American Army has 10,000 of the 12,000 helicopters in US service demonstrates how highly valued they are for army-support missions. Helicopters number about 35% of the total US air strength. Even so, many American planners do not have full confidence in the helicopter during mid-intensity and high-intensity combat; they doubt its ability to survive for any length

American Army OH-58D Scout helicopters equipped with the McDonnell Douglas electronic "eye" called the Mast Mounted Sight. It allows the scout helicopter crew to seek and designate targets at extended ranges in day or night, through haze and smoke. The sight locates targets for attack helicopters and for "smart" and conventional artillery. The helicopter can hover behind trees, buildings and hilly terrain and still give crews a clear view of the target. The system goes into service in 1986 and there are reports that it will be field tested in Central America.

An Israeli Bell 205 helicopter in the south Lebanon security zone, during the 1985 Israeli withdrawal from the positions they had held since 1982. It was used for a variety of purposes, from evacuating wounded to ferrying in supplies.

of time. The helicopter's success in Vietnam, they believe, was due solely to the low-intensity nature of that conflict.

Other planners say that combat helicopters—usually called gunships in American parlance—give commanders the basic requirements of depth of action, versatility of manoeuvre, ability to synchronise parts of an action, and, most important, rapid initiative. US doctrine calls for offensive action to be maintained. Supporters of the helicopter say that it is the most flexible weapon system currently available.

It seems certain that the main American air organisation of the near future will be the combat aviation brigade which will be attached to every infantry division.

Helicopters are becoming more diverse as they assume more roles. The US forces have attack, scout, cargo, utility and medivac helicopters. The first three explain themselves. The utility helicopter provides air mobility for troop assault, troop withdrawal—or "extraction" to the Americans—and supply of troops in combat. The medivac helicopter as a means of getting wounded men quickly out of action and into hospital proved its value in Vietnam. Few of the armies to which the US supplies helicopters have asked for medivac types; this is perhaps more a reflection of high command priorities than of the usefulness of the aircraft.

Since World War II, conflict has been of medium to low intensity. To maintain the low intensity, military thinkers in several countries believe that light forces of extreme flexibility are needed to respond to crises. By just *being there* quickly, a light force can deter further aggression. A conventional heavy force can never be there quickly.

The trend now is towards the use of a multi-purpose light combat helicopter; it does not yet exist in the American forces, but a Light Helicopter Experimental is being developed. The Israeli forces, whose very existence depends on quick response to enemy attack, also sees the need for such an aircraft and might well be the first nation to possess it. Their light combat helicopter will probably also be used in night actions and special operations, such as anti-terrorist warfare.

Vietnam-vintage helicopters could not survive in a high-intensity war; they are too vulnerable to enemy fire. Designers are working on three premises for the ideal helicopter: it must have the means of avoiding detection, if detected, it must have the ability to avoid being hit, if hit, it must be able to withstand damage. Such a helicopter will appear within a few years and to some extent is already here, as with the AH-64 Apache and the UH-60A Black Hawk.

Chemical/Biological Warfare

At least eight establishments in the Soviet Union are researching, producing and storing germ warfare materials. The major factories are at Sverdlovsk, in the Microbiology and Virology Institute, and at Zagorsk, in the Research Institute of Sanitation. Both were built in 1968 and are under heavy military guard. The other six production centres are in the cities of Asku, Berdsk, Kurgan, Omutninsk and Penza.

A special division of the microbiology organisation was created in the early 1970s especially to develop biological weapons. Its chief is General V. I. Ogarkov (no relation to the former Soviet Army commander Marshal Ogarkov).

The Soviet Union is developing a new class of chemical-biological warfare agents that can be rapidly produced for deployment. Some of the compounds developed under the programme are available for extensive field tests. They would then be "weaponised" and coupled with explosive spreader devices.

"Yellow rain" agents (trichothecene mycotoxins) have long been a Soviet chemical warfare agent; they have been used in Afghanistan, Laos and Kampuchea and have been supplied by the Soviet Union to Vietnamese forces. Mycotoxins produce skin blisters, vomiting, diarrhoea and often death. US Intelligence obtained samples of mycotoxins from sites of the attacks, including mycotoxin contamination in a Soviet protective gas mask.

The use of chemical weapons by Iraq in the war against Iran is another trend towards more general biological warfare.

There is a strong case for NATO having some nerve-gas weapons. The latest ones are less worrying, according to specialists, than the existing variety in Western Europe.

The American House of Representatives voted on 19 June 1985 to authorise money for the development of new nerve-gas weapons. As serious commentators have pointed out, it is important for Europeans to think the nerve-gas issue through, so that NATO avoids the type of row which split Europe over neutron bombs in 1977-8.

The Soviet Union has a large stock of nerve gases. If a war broke out and the Russians threatened to use these weapons, NATO's troops would have to fight in their cumbersome gasproof clothing. The Russians would have a tremendous advantage in mobility if they did not have to be similarly handicapped. Should the Soviet Union use its gases, and NATO had none to use, then the Soviet Union would inevitably win. NATO needs a stock of chemical weapons as a deterrent against the Russians using theirs; during World War II both sides were deterred from using gas because both possessed it.

The argument that because NATO has nuclear weapons it does not need chemical ones is illogical. If a monopoly of chemical weapons gave the Soviet the power to win a war in Europe without using nuclear ones, NATO would have a stark choice—Go nuclear or go under. This choice is avoided by preventing the Russians from having a monopoly on chemical war.

The Americans, who provide virtually all of NATO's chemical weapons, need new ones because the present stock is decades out of date; in some cases the material is dangerous. In addition, many weapons which could once have delivered the gases are not now in service.

The chemicals for the new weapons are the same as in many of the old ones, but the new variety will be safer to store and handle. The old ones were ready-mixed and instantly lethal, but the new weapons will be binary; two substances, harmless in themselves, will mix while a shell or bomb is on the way to the target, and only then will be dangerous.

NATO needs just two new projectiles to complete the nerve-gas "systems"—a 155-mm artillery shell and an aircraft bomb.

Leading Western publications concede, however reluctantly, that nerve gas is necessary. An editorial in *The Economist* stated (29 June 1985): "It is a sensible way of tidying up the balance of fear which keeps the peace".

Soviet Test Strike at Turkey

The Soviet Union has successfully tested its armed forces' capability rapidly to invade and conquer Turkey. As a result of the most important military exercise for years, the Soviet command is confident that it can break NATO's southern flank and thus outflank its other defences.

In the exercise, called CAUCASUS 85, the Soviet Army applied tactical lessons learnt in Afghanistan. In fact, as early as spring 1980 the Russians were speculating on the chances of success with a lightning strike at Turkey. Speculation led to reassessments of strategic objectives and the means of achieving them.

CAUCASUS 85 was the major experiment involved. Its essence was to drive a Soviet force through a series of challenging obstacles, from almost impassable terrain to enemy defences strongly held. Great emphasis was put on the functioning of combined arms operations with various small units and sub-units of fighting specialists, such as combat engineers and pioneers. Generals were instructed to experiment with deep outflanking and enveloping manoeuvres.

Sub-unit commanders were told that while they must move rapidly, no competitive race was involved; all the units had to arrive within a short space of time so that their efforts could be co-ordinated. To reach their objectives during CAUCASUS 85, these units had to make their way through defended mountain passes.

The commanders made use of combined operations using ground troops, air-mobile soldiers, helicopters and fighter-bombers. The 6-day period allotted to the CAUCASUS 85 exercise is exactly the time span which the Soviet command believes would be necessary to conquer Turkey.

Soviet strategic interests in the Middle East and Europe have played a great part in the Soviet desire to have only satellite states on Turkey's borders. Thus it will not be necessary to fight through to Turkey.

In Soviet strategic thinking, an Islamic rising in the Muslim states of the Soviet Union is inevitable; it could be forestalled by consolidating Soviet control over portions of eastern Turkey.

But Turkey's membership of NATO is the greatest threat, as the Russians see it. They believe that if they attack Turkey in the context of a *local war*—such as one necessary to check Islamic extremism—the European members of NATO would neither intervene nor follow the war to spread into Central Europe. They are encouraged in this belief by the bitter hostility of Greece, another NATO member, towards Turkey, its neighbour. The result could be that the Soviet Union would achieve its centuries-old desire of unimpeded access to the Mediterranean and the Middle East by land and sea, with all that this would mean in regional strategic and tactical dominance.

CAUCASUS 85, held in blazing summer heat, was the natural successor to

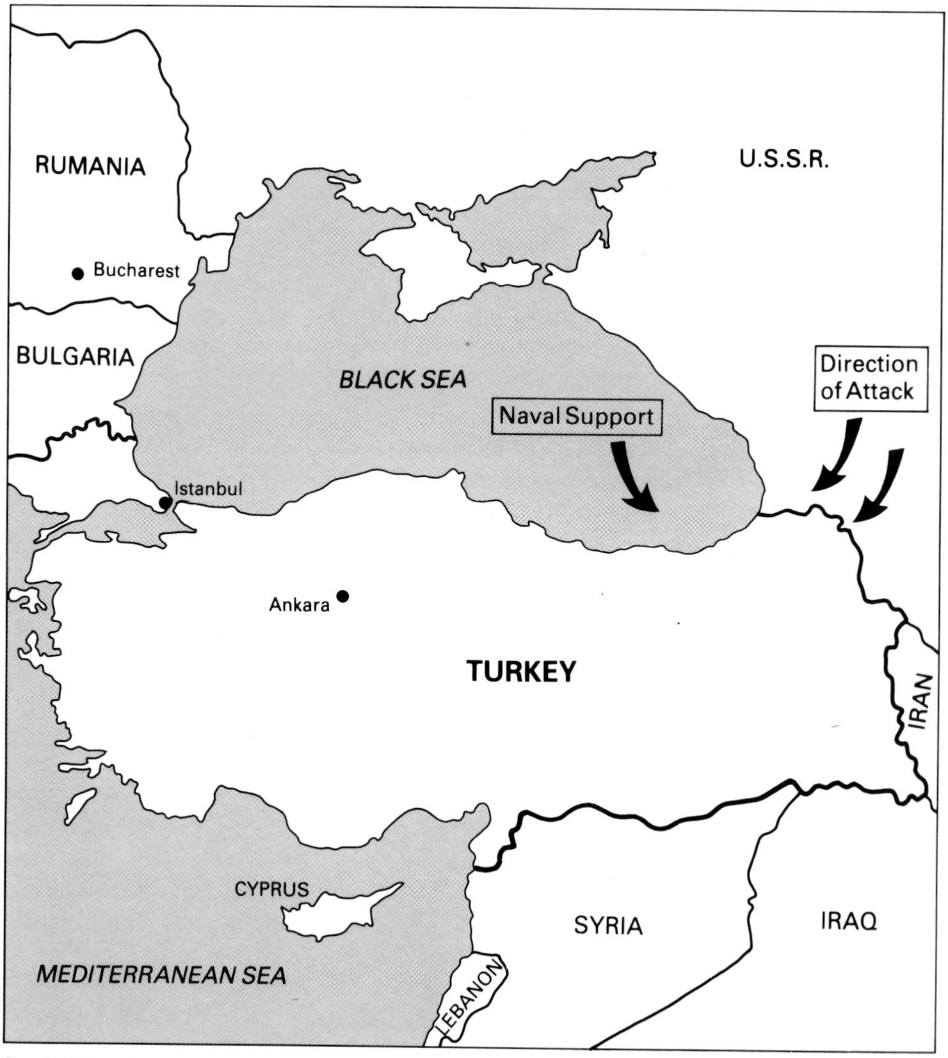

Soviet Test Strike at Turkey

CAUCASUS 76, which was held in mid-winter in southern Georgia. All the lessons from that exercise seem to have been incorporated into CAUCASUS 85.

Certain parts of CAUCASUS 85 are specially significant. Immediately prior to the start of the main battle, Spetsnaz commandos raided "enemy" headquarters. Even though the "enemy" naturally expected such an attack, they were taken by surprise; their command was disrupted at the most critical moment. The Soviet command has great confidence in the ability of Spetsnaz units to achieve the same success against the Turks and other NATO countries.

In addition, the commander of the attacking army rapidly struck with his main force and maintained his momentum, despite strong efforts by the enemy to block or slow him. This is in keeping with the Soviet doctrine of relentless armoured and infantry advance under heavy artillery protection.

The exercise is specially signficant, because some of the Soviet Army's best divisions and commanders were used as enemy. The attackers were not given contrived advantages.

CAUCASUS 85 was carried out in an area 100 km square, south-east of Tiflis and 200 km from the borders of Turkey and Iran. The terrain was precisely the same as that which would have to be negotiated in a strike at either of these countries. Men and equipment were tested to the limit.

Given the right circumstances in Western Europe or the Middle East, the Soviet Union will send its armies into Turkey. The right circumstances could be Turkey's preoccupation with Greek hostility. Already Turkey holds half of its army of 500,000 in the Aegean area and in Thrace, adjacent to Greece and Turkey's toehold in southern Europe.

The teeth of the defence of the NATO's south-eastern flank is the US 6th Fleet—45 warships, 60,000 men and 275 combat aircraft. The main threat is the 5th Squadron of the Soviet Black Sea Fleet, which regularly passes through the Bosphorus and Dardanelles to the Mediterranean. The Turkish high command admits that it cannot defend the straits, in the event of war, without immediate outside reinforcements.

Should Greece withdraw from the NATO defensive ring and terminate the leases of 16 US bases in 1988—as it has threatened to do—then Turkey becomes more vulnerable to the Soviet threat; the Russian commanders will know that the Greek Army will not attack the Soviet flank.

Tensions in the US Pacific Basin

Washington is worrying about its strategic position throughout the Pacific Basin, and the island of Palau has become an even more important part of American military planning. Concern over Palau follows from the chronic instability of the Philippine Islands where the US has spent billions of dollars developing its massive facilities at Clark Air Force Base and Subic Bay Naval Station.

Palau, which was known as Little Tokyo when the Japanese used it as their administrative centre for their Pacific empire of 1941-5, lies at the western end of the Micronesian chain; it has been administered by Washington since 1945.

The US effort to secure a political settlement with the inhabitants of the Palau Republic (15,000 islanders) has been going on since 1975. The impasse is the result of one of the US's most difficult political problems, which began in 1947. The victorious Americans took control of the 11 colonial territories ruled by Germany and Japan and assumed jurisdiction over 2,000 islands. The Palau archipelago is now the only one to remain a colony, and it has been kept sealed off from the rest of the world largely because the Americans designated it a "strategic trust". This allowed the Americans to build military facilities in the region.

The Palau islands were put under the general supervision of the UN Security Council; the General Assembly is responsible for the 10 other trusteeships.

The Palauns have a constitutional ban on all nuclear weaponry, but Palau's great natural harbour is the American alternative to Subic Bay. The US has agreed to subsidise the island for 50 years in return for military facilities. This subsidy amounts to $922.4 million or $1,230 annually for every man, woman and child.

The islanders want close association with the US, but in five separate votes they rejected nuclear weapons. Tensions, acute for years and manifested by violence and bomb attacks on public buildings, reached a climax in July 1985 when President Harue Remelik was assassinated because of his pro-nuclear stance. The Palauns are well aware of what nuclear testing did to Bikini atoll, not so far from Palau.

Similar tensions and fears about being caught up in a nuclear war are spreading to the Commonwealth of the Northern Mariana Islands and the Federated States of Micronesia.

The American position is that in liberating the Pacific Islands from Japanese occupation, they incurred heavy casualties and spent vast amounts of money. They say that they do not want that experience again. Also, they do not want to leave the Pacific open for Soviet expansion outwards from Vietnam.

In addition, secret American strategic analyses warn that Japan has not given up the idea of a Pacific empire. School textbooks are being rewritten to downplay Japanese military and colonial atrocities. After being conquered for the first time in their history, the Japanese suffered a period of self-loathing, one report states, but this is giving way to what Prime Minister Nakasone calls "correct nationalism". Patriotism is being instilled in the new generations.

Nakasone's aim is a renewal of the "Yamato Damashii", the spirit of Japan as the unique country, and the prewar ethics that required a citizen to put duty to state above individual rights. Until he became prime minister in 1983, Nakasone made no secret of his belief that armed with "correct nationalism" and devotion to the state, the Japanese could regain their rightful place as lords of Asia, temporarily ceded to the Americans. Nakasone is neither preparing for war against the US nor advocating war. "Moral transcendance", as he terms it, should be adequate.

As governor of Takushoku University, which before the war trained young men for service in Japan's colonies, Nakasone often reminded students of what he called the "superiority of the Japanese race". He told them: "When we go to Europe we feel the eclipse of the European races. My feeling is that the culture of the coloured races will emerge strongly with Japan as a central force". [The word Takushoku means "settle and exploit".]

The US report quotes a Japanese author, Akiyuki Nosaka, who says that the thrust of Japanese nationalism has always been against the US, ever since American gunboats in the last century forced Japan, then a closed country, to open itself to trade with the rest of the world. The American conquest and occupation after 1945 aggravated what Nosaka calls "our continuing vendetta".

Nosaka predicts intense competition between the two nations for economic influence in China, and resentment at American opposition to Japan's having nuclear weapons. "Japan's urge to dominate is getting stronger", says Nosaka. "Japan is a very dangerous country".

Prime Minister Nakasone strongly believes that Japan can only command international respect through increased firepower. Under Nakasone, the campaign for the return of four islands held by the Soviet Union has grown more strident. Japan has asked the US to indicate Japan's claim to the islands on new maps.

Militarism was rendered impotent by the "peace constitution" imposed by the Americans after 1945, but it is showing itself again and confidence is high in the armed forces. A former prime minister, Masayoshi Chira, in 1980 made the statement that: "The United States is no longer a superpower". It is this belief which leads Japanese strategic planners to the conclusion that the Pacific, and then wider fields, will become Japan's preserve.

Role Reversal of the Superpowers

One of the most significant and interesting war trends is that in many civil wars the Soviet-backed regimes face insurgencies while the US supports the rebels.

This is in direct contrast to a decade ago when the Russians were expanding their empire and their influence by inciting and supporting leftist insurgencies. In that period the Americans were backing governments. Some of the current wars illustrate the reversal of roles.

Nicaragua. In helping the Contras against the Sandanista government, the US is quite openly waging what many Americans prefer to call a "covert" operation.

Afghanistan. Shoring up a puppet regime, the Soviet Army wages a ferocious pacification campaign against the rebel tribesmen with helicopter gunships, napalm strikes and carpet bombing.

Angola. The pro-Soviet government relies heavily on 30,000 Cuban troops—who are also part of the Soviet camp—as well as on immense amounts of Russian materiel. It is only a half-kept secret that the US supports UNITA, the rebel force.

Ethiopia. The Soviet Union backs and props up yet another regime while the US supports the guerrilla rebels fighting against it.

In effect, the Russians are defending the established order—that is, they are anti-revolution—while the Americans, through their clients, are challenging the established order. This could well be the result of a mistaken Moscow strategy which over-extended its expansionism and left itself open to Western counter-attack in the mid and late 1980s. The Americans are tending towards covert action of the 1960s type, going on the offensive to put the Russians on the defensive. In itself, this is a reversal of roles.

By direct action, rather than through unreliable, unstable or fearful allies, the US has an unannounced strategy of action behind enemy lines. There are few conflicts in the world today where they are not doing this in one form or another. "Support for freedom fighters is self-defence", President Reagan has said. This kind of assertive deterrence is an effective way of keeping the Soviet Union contained.

The problem is to know how far to go in supporting anti-Communist insurgencies. Sometimes the rebels are morally no better than their enemies of the regime. The most notorious example is Pol Pot, a leader of the anti-Vietnamese resistance—and therefore anti-Soviet—who sanctioned the murder of probably 2 million of his own countrymen.

The switch of roles with the Russians gives the Americans options it did not have when it was backing governments. For instance, the Reagan administration, having deliberately pursued a policy of fight *and* talk in Nicaragua—backing the Contras while working for negotiations with the Sandanistas—just as deliberately moved to a policy of fight *or* talk. This was a tacit invitation to the Sandanista government to get itself off the hook by agreeing to serious negotiations.

The American public and the US congress is not ready for unlimited secret war, but the trend is in that direction.

Insurgencies in the 1970s

Country	*Type*
Angola	Nationalist/leftist
Ethiopia	Ethnic/nationalist
Kurds (Iraq)	Ethnic/rightist
Mozambique	Nationalist/leftist
Oman (Dhofar)	Nationalist/communist
South-East Asia	Communist
Western Sahara	Nationalist/leftist

Insurgencies in 1985

Country	*Type*
Afghanistan	Islamic/nationalist/rightist
Angola	Ethnic/nationalist/rightist
Chad	Ethnic/Islamic (Gaddafi)
El Salvador	Nationalist/rightist
Kampuchea (formerly Cambodia)	Nationalist/anti-Vietnamese
Lebanon	Islamic/leftist
Mozambique	Ethnic/nationalist/rightist
Nicaragua	Nationalist/rightist
Peru	Independent Maoist
Philippines	Ethnic/religious/leftist
Sri Lanka	Ethnic
Sudan	Ethnic
Uganda	Ethnic
Western Sahara	Nationalist/leftist

Six of the insurgencies listed above have a strong sectarian element. They are Afghanistan; Chad; Lebanon; Philippines; Sri Lanka; Sudan. Apart from these six another eight of the wars described in this book are, to a degree, "religious" wars. They are: Bangladesh; Burma; East Timor; the Gulf War; India-Kashmir; Northern Ireland; Sikh-Hindu war. Jihad, of course, is essentially a religious war.

South Africa—A Short Cut to War?

Suddenly a "popular rising" or revolution in South African does not seem as far off as it did in 1984. The African National Congress (ANC), the major South African liberation movement, long convinced that only a protracted guerrilla war could overthrow the white regime, is encouraging its supporters in South Africa to work towards an Iranian-style insurrection.

About 350 people were killed during 1985 in the continual disturbances in black townships; many of the victims were black policemen and town councillors killed by their own black fellow citizens because of their collaboration with the white regime. There were also brief but successful general strikes by black workers in industrial areas of the Transvaal and Eastern Cape Province.

The ANC called not only for a long general strike but for armed bands of young people to turn the townships into "no-go areas" for anything short of a very large force of police or soldiers.

An ANC spokesman in Lusaka, Zambia, where the organisation has its headquarters, said in mid-1985:

> "Our call to arms contains new and far-reaching objectives which we believe are now realisable because of the astonishing speed with which the situation is changing. The idea of a general insurrection as an immediate way forward cannot replace the long-term perspective we have of a people's war, but we should keep the lines open to other possibilities. The possibility of bringing about the collapse of the present set-up in South Africa through the build up of insurrectionary forces has never been as great as it is today".

Despite the spokesman's confidence, South Africa is not at the brink of an insurrection which could become a "popular rising". And should this happen, it is not likely to be successful. All but a couple of casualties have been blacks, and it is still mostly the police—not the army—who are used to contain them. (South Africa has only 1.4 policemen for every 1,000 members of the public compared with 2.4 in Britain.)

The model of the Iranian revolution is inapplicable; in Iran practically the entire nation was against the shah, his soldiers and his police. In South Africa it is one community against another, and the white community has nearly all the firearms.

However, even a failed insurrection could bring about the transformation of South Africa, which must occur before the ANC really starts to fight. For instance, the townships and homelands must be able to provide the ANC with the guerrilla base areas that it cannot hope to have across the borders because of South Africa's ability to inflict huge military retaliation on its neighbours.

It could be a decade or more before the "protracted people's war" might bring down the white regime, but the ANC has always accepted this. Its problem until now has been just to get the war properly started. Until 1985 even that looked like

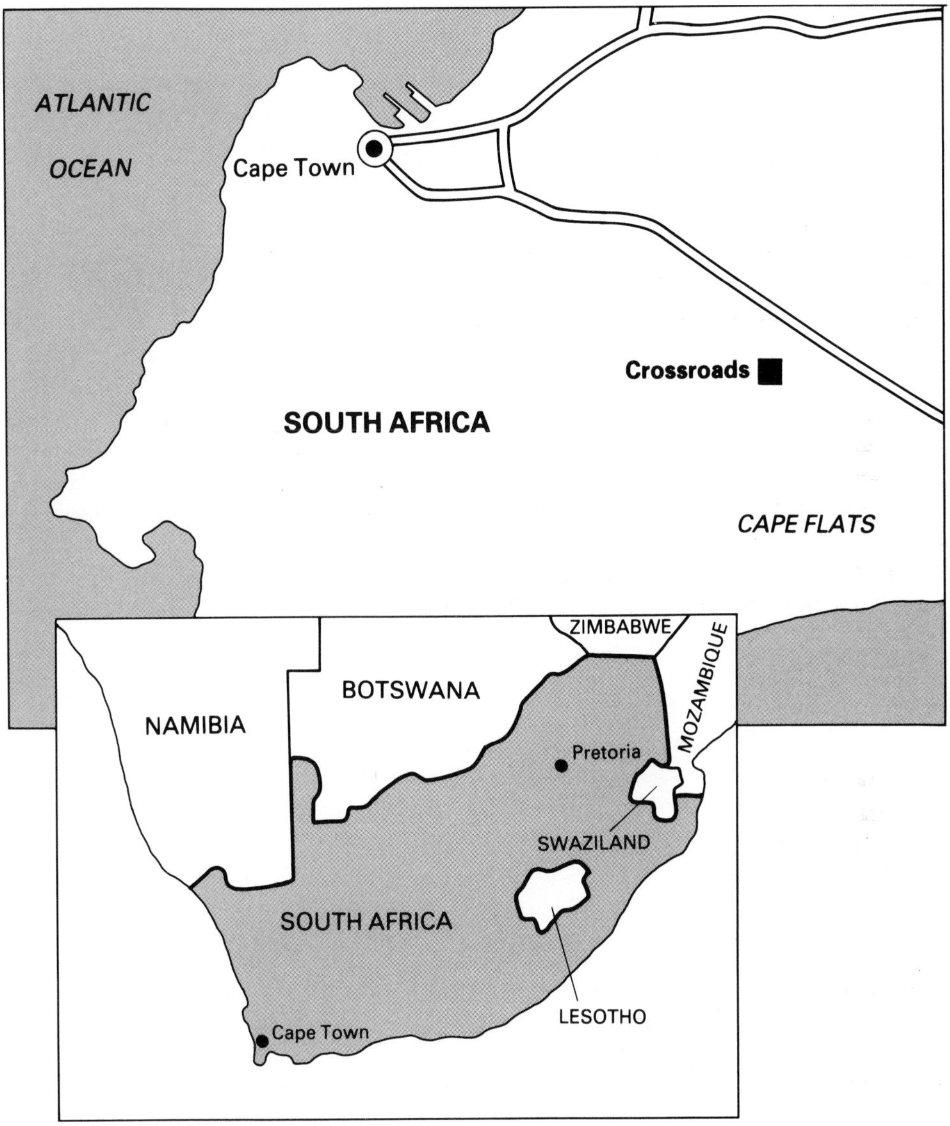

A Short Cut to War – South Africa

a task that could take years. It might still do so, but the ANC high command is now hoping that a real insurrection in South Africa could be the short cut to war. This trend is one of the most significant of the present year.

This trend, one of the most significant of the year, continued in 1986 with almost continuous violence. On several occasions ANC terrorist squads exploded bombs in white areas. A bomb at a police headquarters in Johannesburg demonstrated the ANC's increasing aggressiveness and daring. Whites were attacked in their cars and in their homes, as were black policemen, several of whom were murdered. Several other blacks allegedly guilty of "collaboration" with the South African security forces were killed by "necklaces"—tyres filled with petrol, put around the victim's neck and set alight.

In dispersing demonstrators the police used firearms and tear gas even at funerals, in some incidents as many as 10 blacks were killed, thus further increasing black anger.

In 1986 political leadership seemed to be crystallising around Mrs. Winnie Mandela, wife of the black leader Nelson Mandela, who had been in prison for 25 years. Mrs. Mandela, in common with Bishop Desmond Tutu, called for international economic sanctions against South Africa and strongly attacked the United States and Great Britain for failing to apply sanctions. The "short cut to war" appeared to clearly marked by the summer of 1986.

By the same author

Military

Middle East Journey
Return to Glory
One Man's War
The Walking Wounded
Digger (The Story of the Australian Soldier)
Scotland the Brave (The Story of the Scottish Soldier)
Jackboot (The Story of the German Soldier)
Tommy Atkins (The Story of the English Soldier)
Jack Tar (The Story of the English Seaman)
Swifter than Eagles (Biography of Marshal of the R.A.F. Sir John Salmond)
The Face of War
British Campaign Medals
Codes and Ciphers
Boys in Battle
Women in Battle
Anzacs at War
Links of Leadership (Thirty Centuries of Command)
Surgeons in the Field
Americans in Battle
Letters from the Front 1914-18
The French Foreign Legion
Damn the Dardanelles! (The Story of Gallipoli)
The Australian Army at War 1899–1975
The Israeli Army in the Middle East Wars 1948–1973
The Arab Armies in the Middle East Wars 1948–1973
Fight for the Falklands!
On the Western Front: Soldiers' Stories 1914–18
The Man the Nazis Couldn't Catch
The War of Desperation: Lebanon 1982-1985
Brassey's Battles—3,500 years of conflict, campaigns and wars from A–Z

General

The Hunger to Come (Food and Population Crises)
New Geography 1966-67
New Geography 1968-69
New Geography 1970-71
Anatomy of Captivity (Political Prisoners)
Devil's Goad
Fedayeen (The Arab-Israeli Dilemma)
The Arab Mind
The Israeli Mind
The Dagger of Islam
The PLO Connections
The Arabs as Master Slavers
Know the Middle East

And other titles

N. Ireland
Grenada
W. Sahara
Honduras
Guatemala
El Salvador
Nicaragua
Colombia
Ecuador
Peru
Chile